SpringerBriefs in Ecology

For further volumes:
http://www.springer.com/series/10157

Claudio O. Delang · Wing Man Li

Ecological Succession on Fallowed Shifting Cultivation Fields

A Review of the Literature

 Springer

Claudio O. Delang
Department of Geography
Hong Kong Baptist University
Kowloon Tong
Hong Kong SAR

Wing Man Li
Shatin, New Territories
Hong Kong SAR

ISSN 2192-4759   ISSN 2192-4767 (electronic)
ISBN 978-94-007-5820-9   ISBN 978-94-007-5821-6 (eBook)
DOI 10.1007/978-94-007-5821-6
Springer Dordrecht Heidelberg New York London

Library of Congress Control Number: 2012954055

Printed on acid-free paper

Springer is part of Springer Science+Business Media (www.springer.com)

Acknowledgments

The work described in this book was fully supported by a grant from the Research Grants Council of the Hong Kong Special Administrative Region, China (project no. HKBU442609).

Contents

1 **Introduction**   1
   References   5

2 **Forest Structure**   9
  2.1 Introduction   9
  2.2 Total Aboveground Biomass   10
     2.2.1 Aboveground Biomass Increases with Fallow Age   10
     2.2.2 Total Aboveground Biomass Increases at First, but
          then Remains Constant   11
  2.3 Trees and Herbs Abundance   12
  2.4 Total Basal Area   15
  2.5 Canopy Height   17
     2.5.1 Canopy Height Increases with Fallow Age   17
     2.5.2 No Significant Difference with Fallow Age   18
  2.6 Plant Density   19
     2.6.1 Stem Density Increases, or Increases
          and then Stabilizes   19
     2.6.2 Stem Density Increases, but Shows a Decline
          in Early or Intermediate Stages   21
     2.6.3 Stem Density Varies Among Age Classes   22
     2.6.4 Stem Density Decreases with Fallow Age   23
  2.7 Discussion and Conclusions   25
  References   35

3 **Species Richness and Diversity**   39
  3.1 Introduction   39
  3.2 Total Species Richness   40
     3.2.1 Species Richness Increases with Fallow Age   41
     3.2.2 Species Richness Increases at First, then Drops Off
          or Stays Constant   48
     3.2.3 No Significant Difference Between Age Classes   51

3.3   Tree and Herb Species Richness   51
      3.3.1   Species Richness Increases with Fallow Age   51
      3.3.2   Species Richness Peaks at Early to Intermediate Stages. . .   52
3.4   Total Species Diversity   54
      3.4.1   Species Diversity Increases Rapidly,
              then Remains Constant   54
      3.4.2   Species Diversity Increases with Fallow Age   56
      3.4.3   Species Diversity Decreases When Fallows Age.   62
3.5   Discussion and Conclusions   62
References   65

4   Species Composition.   67
    4.1   Introduction   67
    4.2   Succession from Grasses to Trees   69
          4.2.1   Location-Specific Patterns of Change   70
    4.3   Fallow Length and Species Composition   71
          4.3.1   Similarity of Species Composition
                  in Similar-Aged Fallows   73
          4.3.2   Pioneer Species Recruit First During Succession.   75
          4.3.3   Both Pioneer and Forest Species Recruit Together
                  in the Early Stage of Succession.   77
          4.3.4   Replacement of Pioneers by Shade-Tolerant Species. . . .   80
          4.3.5   Time to Reach Similar Species Composition
                  as the Old-Growth Forest.   82
          4.3.6   High Variability of Species Composition   83
    4.4   Discussion and Conclusions   86
    References   88

5   Factors Contributing to Differences in Forest Recovery Rates   91
    5.1   Introduction   91
    5.2   Number of Slash-and-Burn Cycles.   92
    5.3   Length of Fallow Period.   98
    5.4   Type of Soil   103
          5.4.1   Faster Forest Regeneration in Soil with Higher Fertility
                  and Better Structure   104
          5.4.2   Soil Properties Have No Effect on Forest Structure.   110
          5.4.3   Impact of Shortened Fallow Duration on Soil Fertility . . .   110
    5.5   Types of Forest.   115
    5.6   Discussion and Conclusions   116
    References   118

6   Conclusions   123
    References   126

Chapter 1
Introduction

Keywords Chronosequence method · Permanent plot method · Political constraints · Soil erosion · Watershed degradation · Logging · Agricultural intensification · Hydroelectricity production · Nature protection

For thousands of years, swiddening, shifting cultivation, or slash-and-burn farming, as it is alternatively called, has been a common farming practice (Jääts et al. 2011). While it has been largely replaced in temperate countries by more sustainable agricultural practices, it continues in many tropical countries (van Vliet et al. 2012). The reasons for continued practice are varied. From a socioeconomic point of view, they include the low population density in the area in which the farmers live, which allows them to use shifting cultivation (Delang 2006a), and the lack of access to permanent fields, either for legal or for economic reasons, which forces them to use shifting cultivation (Ochoa-Gaona 2001). From an ecological perspective, poor soil quality is a major factor. In tropical countries, much of the nutrients necessary to grow crops are in the vegetation rather than the soil, and are released in the form of ashes when trees are burned (Neyra-Cabatac et al. 2012). After the land is cleared, it may be cultivated for a few years without using pesticides or fertilizers. However, the land quickly degrades, and without the use of fertilizers and pesticides the yield is soon very low and the farmers have to move on, clearing new land.

Because the swiddens have to be left fallow for several years after only a few years of cropping, slash-and-burn farming requires large amounts of forested land in order to be sustainable. For this reason, a large number of critics, from colonial powers (see Scott 2011) to the FAO (Brown 1997), and from academics (Ickis and Rivera 1997) to the popular press (Bangkok Post 1999), blame swiddening for what they consider to be an "inefficient use of the forest", which ultimately leads to deforestation (Henley 2011; Ellen 2012), soil erosion (Fullen et al. 2011), and the degradation of the watersheds (Ayanu et al. 2011), which is largely responsible for the water imbalance (Ziegler et al. 2004; Ellen 2012). The alternatives

C. O. Delang and W. M. Li, *Ecological Succession on Fallowed Shifting Cultivation Fields*, SpringerBriefs in Ecology, DOI: 10.1007/978-94-007-5821-6_1, © The Author(s) 2013

proposed include "more efficient uses" of the land, such as logging and the export of wood, which earn a country foreign currency (Ekoko 2000), agricultural intensification and the production of exportable cash crops, construction of dams to produce hydroelectricity (Delang and Toro 2011), or forest conservation in the name of nature protection. Some of these "more efficient uses" are not intended to protect the forest per se, but simply aim to generate higher returns than swiddening does. The problem of redistribution of these higher returns, and in particular how much swiddeners should receive as compensation for their loss of land, is rarely addressed, and when it is, it is dealt with unsatisfactorily (Delang 2006b).

Since land degradation is used as the major justification for outlawing swiddening, fundamental questions remain regarding the nature of ecological succession on fallow areas following field abandonment. For example, how long does land need to remain fallow before it can be farmed again? What kind of plants can people gather during the fallow period when the vegetation re-grows? How does species diversity, richness and composition change in fallowed areas as the secondary forest re-grows? How does vegetation density and abundance change? These are essential questions that should direct government policies towards shifting cultivation. Yet, they are not sufficiently addressed, partly because of the lack of voice of the shifting cultivators, and therefore disinterest by government officials, and partly because there is insufficient knowledge of ecological succession.

One problem is that the results of research on ecological succession in a particular region may not be easily extrapolated to other sites, which harbour different ecosystems or biomes. This problem can be addressed through an extensive review of the literature. Although small meta-analyses have been carried out, there is a need for a more extensive literature review. This book addresses that need by reviewing the literature on the ecological succession of fallowed areas following shifting cultivation. The literature is divided into four chapters/main themes, each of which focuses on a particular issue. Case studies are selected from tropical and subtropical forests in Central and South America, Africa and Asia (Fig. 1.1).

In this book, we look at ecological succession as a natural process wherein vegetation re-grows after disturbance by humans, mainly for agricultural purposes (Faber-Langendoen 1992; Guariguata and Ostertag 2001). Changes in the structural characteristics of forests, such as biomass, basal area, canopy height, species richness, and species composition are the most commonly measured characteristics of forests when examining ecological succession (Guariguata and Ostertag 2001). This book discusses how these characteristics change as the fallows age.

Broadly speaking, there are two approaches to examine the process of ecological succession: chronosequence studies and permanent plot studies (Chazdon et al. 2007; Quesada et al. 2009). Chronosequence studies select a number of fallow stands with different ages in order to explore the rate of vegetation change over time (Chazdon et al. 2007; Lehmkuhl et al. 2003; Kalacska et al. 2005). Although chronosequence study is one of the most commonly used methods to examine the succession process following shifting cultivation (Aide et al. 2000; Breugel et al. 2007; Chazdon et al. 2007; Trejos-Lebrija and Bongers 2008;

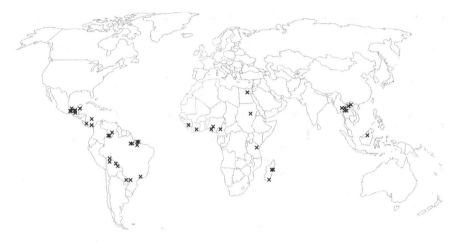

Fig. 1.1 Location of the sites reviewed in this book

Tran et al. 2011), it has been criticized for its bias. Quesada et al. (2009) reviewed a few criticisms. Chronosequence studies are based on the assumption that sites represent a continuum. However, Quesada et al. pointed out that this assumption is often not verified by researchers. To assume constant biotic and abiotic conditions at the study site may be inaccurate, as such conditions might change over time (Johnson and Miyanishi 2008). Moreover, Quesada et al. pointed out that as chronosequence studies do not directly examine the vegetation changes along a time sequence, it should be implemented carefully, as bias would inevitably arise (for example, if an old-growth forest is used to do comparisons, it should show the same variations in soil and topography as the fallow stands).

The permanent plot method involves directly monitoring sites for a long period of time to see how vegetation communities change, in other words "directly [documenting] the rates of change through monitoring vegetation dynamics over time in particular forest stands" (Chazdon et al. 2007, p. 274). By doing this, scientists may observe successional changes on the same site over a long period of time (Sheil et al. 2000; Sheil 2001; Chazdon et al. 2007). A drawback of this approach is that to carry out a research on a large scale involves a great amount of human and financial resources (Quesada et al. 2009). Thus, understandably, the number of plots that can be surveyed is limited. Furthermore, the processes observed in one location might not reflect those found in other geographic regions: as Duivenvoorden (1995), Givnish (1999) and Balvanera et al. (2002) pointed out, tropical areas have an heterogeneous environment.

Given these problems, Chazdon et al. (2007) concluded that chronosequence studies are easier to carry out and may consider a longer time frame. Moreover, they suggested that while climatic variations might appear between years, by using chronosequence studies, these variations can be corrected. Despite its limitations, Quesada et al. (2009) concluded that chronosequence studies are better, as they can overcome the restrictions of studying permanent plots. Also, as pointed out by

Quesada et al. (2009), most studies can minimize biases by selecting sites with similar soil type, topography and land-use history (e.g. Lebrija-Trejos et al. 2008; Raharimalala et al. 2010; Tran et al. 2011; Saldarriaga et al. 1988; Metzger 2003; Kammesheidt 1998).

In the following chapter (Chap. 2) we review the literature on forest structure. There is no overall measure to quantify or express forest structure, and foresters measure a variety of attributes—aboveground biomass, abundance, basal area, canopy height, plant density—each of which contribute to the overall structure, but do not, individually, describe it completely. Studies reported great variations in the time needed for the various attributes to reach old-growth forest levels, which implies that the conditions or characteristics of the forest plays a part in altering the regeneration of forest structure during succession. Authors also highlighted the importance of local species composition on influencing the changes in vegetation, which is seldom mentioned or examined in studies of forest succession.

In Chap. 3 we discuss species richness and species diversity. The two concepts are closely related, but are not synonyms. Species richness can be expressed numerically through an account of the number of species: species richness is estimated dividing the number of species by the geographical area. On the other hand, species diversity is a function of the number of species present (i.e. species richness or number of species) and the evenness with which the individuals are distributed among these species (i.e. species evenness, species equitability, or abundance of each species). In this chapter, we divide the literature under the subheading 'total species richness', 'tree and herb species richness', and 'total species diversity'. The review of the literature reveals that some studies observed an increase in species richness and species diversity when fallows age, while others found an increase at the beginning but then a drop in levels (or constant levels) in later stages. A few studies even concluded that there is no difference among age classes. Therefore, we are not able extrapolate a general model for how total species richness is impacted by fallow duration. Similarly, the time needed for total species richness to recover to levels found in old-growth forest also shows great variations.

In Chap. 4 we review research on species composition: how vegetation shifts from grasses to trees in different successional stages. It is generally accepted that a transition from grasses to herbs to shrubs to woody species is common, despite the fact that time of recovery may vary in different places. Also, it is usually suggested that with a shorter fallow, tree growth would be hindered, and the land might not be able to reach a tree fallow. We also discuss what different researches have found regarding the relationship between soil depletion and the reduction of tree species as fallow duration shortens. Whether the changes in species composition are predictable is one of the major questions regarding secondary forest succession. Two influential models have been proposed to predict floristic change: the 'relay floristics' model and the 'initial floristic composition' model. However, our review concludes that new approaches are needed to explain the great variability of species composition among sites, as neither the 'reductionist' nor the 'holistic' approaches seem to be able to fully predict and explain the regeneration pathway of species composition along the successional trajectory.

Chapters 3 and 4 conclude that the studies reviewed found considerable differences in the changes in species richness and diversity, plant density, vegetation abundance and species composition in fallows during succession. These differences imply that forest regeneration is determined by a complex set of factors besides fallow duration. In Chap. 5 we review these factors: the number of slash-and-burn cycles (the number of times the fields were burned), fallow length, the type of soil, and the type of forest. Most of the research on this topic has concluded that these factors influence forest regeneration processes: fewer slash-and-burn cycles, lower shifting cultivation intensity, and more fertile soils leads to greater biomass accumulation, higher plant density, higher tree height and larger basal area. However, concerning soil fertility, other studies found no difference in species richness between more fertile and less fertile soils despite the fact that species composition showed a distinct pattern on different soils. How the ecological succession process following shifting cultivation is influenced by soil properties needs more research and attention, as not much research has been conducted to examine the effect of soil fertility and texture on vegetation changes.

Chapter 6 concludes the book, with a brief summary of the main findings, and the implications and recommendations for further research.

References

Aide TM, Zimmerman JK, Pascerella JB, Rivera L, Marcano-Vega H (2000) Forest Regeneration in a chronosequence of tropical abandoned pastures: implications for retoration ecology. Restor Ecol 8:328–338

Ayanu YZ, Nguyen TT, Marohn C, Koellner T (2011) Crop production versus surface-water regulation: assessing tradeoffs for land-use scenarios in the Tat Hamlet Watershed, Vietnam. Int J Biodivers Sci Ecosyst Services Manag 7(3):231–244

Balvanera P, Lott E, Segura G, Siebe C, Islas A (2002) Patterns of β-diversity in a Mexican tropical dry forest. J Veg Sci 13:145–158

Bangkok Post (1999) Natural scapegoats. By Supara Janchitfah. 6 June

Breugel MV, Bongers F, Martínez-Ramos M (2007) Species dynamics during early secondary forest succession: recruitment, mortality and species turnover. Biotropica 35:610–619

Brown C (1997) Regional study, The South Pacific. Rome: FAO Forestry Planning and Statistics Branch

Chazdon RL, Letcher SG, van Breugel M, Martínez-Ramos M, Bongers F, Finegan B (2007) Rates of change in tree communities of secondary neotropical forests following major disturbances. Philos Trans R Soc B 362:273–289

Delang CO (2006a) Not just minor forest products: the economic rationale for the consumption of wild food plants by subsistence farmers. Ecol Econ 59:64–73

Delang CO (2006b) Indigenous systems of forest classification: understanding land use patterns and the role of NTFPS in shifting cultivator's subsistence economies. Environ Manage 37(4):470–486

Delang CO, Toro M (2011) Hydropower induced displacement and resettlement in Lao PDR. South East Asia Res 19(3):567–594

Duivenvoorden J (1995) Tree species composition and rain forest-environment relationships in the middle Caquetá area, Colombia, NW Amazonia. Vegetatio 120:91–113

Ekoko F (2000) Balancing politics, economics and conservation: The case of the Cameroon forestry law reform. Dev Change 31(1):131–154

Ellen R (2012) Studies of swidden agriculture in Southeast Asia since 1960: an overview and commentary on recent research and syntheses. Asia Pacific World 3(1):18–38

Faber-Langendoen D (1992) Ecological constraints on rain forest management at Bajo Calima, Western Colombia. Forest Ecol Manage 53(1):213–244

Fullen MA, Booth CA, Panomtaranichagul M, Subedi M, Mei LY (2011) Agro-environmental lessons from the 'sustainable highland agriculture in South-East Asia' (SHASEA) project. J Environ Eng Landscape Manage 19(1):101–113

Givnish T (1999) On the causes of gradients in tropical tree diversity. J Ecol 87:193–210

Guariguata MR, Ostertag R (2001) Neotropical secondary forest succession: changes in structural and functional characteristics. For Ecol Manage 148:185–206

Henley D (2011) Swidden farming as an agent of environmental change: ecological myth and historical reality in Indonesia. Environ History 17(4):525–554

Ickis JC, Rivera J (1997) Cerro cahui. J Bus Res 38(1):47–56

Jääts L, Konsa M, Kihno K, Tomson P (2011) Fire cultivation in Estonian cultural landscapes. In: Peil T (ed) The space of culture—the place of nature in estonia and beyond. Tartu University Press, Tartu (Ch. 1)

Johnson E, Miyanishi K (2008) Testing the assumptions of chronosequences in succession. Ecol Lett 11(5):419–431

Kalacska MER, Sanchez-Azofeifa GA, Calvo-Alvarado JC, Rivard B, Quesada M (2005) Effects of season and successional stage on Leaf Area Index and spectral vegetation indices in three mesoamerican tropical dry forests. Biotropica 37:486–496

Kammesheidt L (1998) The role of tree sprouts in the restoration of stand structureand species diversity in tropical moist forest after slash-and-burn agriculture in EasternParaguay. Plant Ecology 139(2):155–165

Lebrija-Trejos E, Bongers F, Pérez-García E, Meave J (2008) Successional change and resilience of a very dry tropical deciduous forest following shifting agriculture. Biotropica 40:422–431

Lehmkuhl J, Everett R, Schellhaas R, Ohlson P, Keenum D, Riesterer H, Spurbeck D (2003) Cavities in Snags along a wildfire chronosequence in Eastern Washington. J Wildl Manag 67:219–228

Metzger JP (2003) Effects of slash-and-burn fallow periods on landscape structure. Environ Conserv 30(4):325–333

Neyra-Cabatac NM, Pulhin JM, Cabanilla DB (2012) Indigenous agroforestry in a changing context: The case of the Erumanen ne Menuvu in Southern Philippines. Forest Policy Econ 22:18–27

Ochoa-Gaona S (2001) Traditional land-use systems and patterns of forest fragmentation in the highlands of Chiapas, Mexico. Environ Manage 27(4):571–586

Quesada M, Sanchez-Azofeifa GA, Alvarez-Anorve M, Stoner KE, Avila-Cabadilla L, Calvo-Alvarado J, Castillo A, Espírito-Santo MM, Fagundes M, Fernandes GW, Gamon J, Lopezaraiza-Mikel M, Lawrence D, Morellato LPC, Powers JS, Neves F de S, Rosas-Guerrero V, Sayago R, Sanchez-Montoya G (2009) Succession and management of tropical dry forests in the Americas: review and new perspectives. Forest Ecol Manage 258:1014–1024

Raharimalala O, Buttler A, Ramohavelo CD, Razanaka S, Sorg JP, Gobat JM (2010) Soil-vegetation patterns in secondary slash and burn successions in Central Menabe, Madagascar. Agric Ecosyst Environ 139:150–158

Saldarriaga JG, West DC, Tharp ML, Uhl C (1988) Long-term chronosequence of forest succession in the upper Rio Negro of Colombia and Venezuela. J Ecol 76:938–958

Scott JC (2011) The art of not being governed: an anarchist history of upland southeast Asia. Yale University Press, New Haven

Sheil D (2001) Long-term observations of rain forest succession, tree diversity and responses to disturbance. Plant Ecol 155:183–199

Sheil D, Jennings S, Savill P (2000) Long-term permanent plot observations of vegetation dynamics in Budongo, a Ugandan rain forest. J Trop Ecol 16:765–800

Tran VD, Osawa A, Nguyen TT, Nguyen BV, Bui TH, Cam QK, Le TT, Diep XT (2011) Population changes of early successional forest species after shifting cultivation in Northwestern Vietnam. New Forest 41:247–262

Trejos-Lebrija E, Bongers F (2008) Succession change and resilience of a very dry tropical deciduous forest following shifting agriculture. Biotropica 40:422–431

van Vliet N et al (2012) Trends, drivers and impacts of changes in swidden cultivation in tropical forest-agriculture frontiers: a global assessment. Global Environ Change 22(2):418–429

Ziegler AD, Giambelluca TW, Tran LT, Vana TT, Nullet MA, Fox J, Vien TD, Pinthong J, Maxwell JF, Evett S (2004) Hydrological consequences of landscape fragmentation in mountainous northern Vietnam: evidence of accelerated overland flow generation. J Hydrol 287(1–4):124–146

Chapter 2
Forest Structure

Abstract This chapter reviews the literature on forest structure. There is no overall measure to quantify or express forest structure, and foresters measure a variety of attributes—aboveground biomass, abundance, basal area, canopy height, plant density—each of which contribute to the overall structure, but do not, individually, describe it completely. Studies reported great variations in the time needed for the various attributes to reach old-growth forest levels, which implies that the conditions or characteristics of the forest play a part in altering the regeneration of forest structure during succession. Authors also highlighted the importance of local species composition on influencing the changes in vegetation, which is seldom mentioned or examined in studies of forest succession.

Keywords Forest structure · Biomass · Trees abundance · Herbs abundance · Basal area · Canopy height · Plant density

2.1 Introduction

There is no overall measure to quantify or express forest structure, and foresters measure a variety of attributes—aboveground biomass, abundance, basal area, canopy height, plant density—each of which contribute to the overall structure, but do not, individually, describe it completely. This is well reflected in the dictionary definition of structure: "a complex system considered from the point of view of the whole rather than of any single part". It is meaningless to simply add together the various measures, and produce some average quantification of forest structure. Hence, in this chapter we review the literature on changes in total aboveground biomass, total basal area, tree and herb abundance, canopy height and plant density in separate sections.

C. O. Delang and W. M. Li, *Ecological Succession on Fallowed Shifting Cultivation Fields*, SpringerBriefs in Ecology, DOI: 10.1007/978-94-007-5821-6_2, © The Author(s) 2013

2.2 Total Aboveground Biomass

To estimate the changes in abundance of vegetation along ecological succession processes, total aboveground biomass and total basal area are commonly measured. Both measurements are calculated from stem diameters, thus are not independent of each other (Chazdon et al. 2007; Clark and Clark 2000). However, many think that aboveground biomass more accurately reflects the conditions of the study site (Chazdon et al. 2007; Brown 1997). Studies on the aboveground biomass have reached different conclusions. Some studies have shown that aboveground biomass increases with fallow age, while others have found that total aboveground biomass increases at first, but then remain constant. The two cases are now reviewed in turn.

2.2.1 Aboveground Biomass Increases with Fallow Age

In the mixed deciduous forest of central Burma, Fukushima et al. (2007) found a constant increase in total aboveground biomass (including herbs and grass) in relation to fallow length. The authors adopted the allometric equations of Ogawa et al. (1965) for the calculation of tree biomass:

$$W_s = 0.0396 (DBH^2 H)^{0.9326}, \ (kg, \ cm^2 m)$$
$$W_b = 0.00602 (DBH^2 H)^{1.027}, \ (kg, \ cm^2 m)$$
$$1/W_l = 26/(W_s + W_b), \ (kg, \ kg)$$

where W_s is the dry weight of the stem, W_b is the dry weight of branches and W_l is the dry weight of the leaves of a tree. For herbs and grasses, samples were collected and weighed after drying in an oven. Fukushima et al. (2007) reported that the total aboveground biomass contained 6.6 t/ha in the first year of fallow, then increased to 55.0 t/ha in the fifth year, and eventually reached 150.0 t/ha in year 40. 43.3–52.9 % of the total aboveground biomass of the old-growth forest was attained between 10 and 18 years after abandonment, and increased to 93.6 % approximately 40 years after abandonment.

Studies carried out in the upper Rio Negro (Uhl 1987), northern Thailand (Fukushima et al. 2008) and the southern Yucatan Peninsular region of Mexico (Read and Lawrence 2003) also found a general trend towards increases in aboveground biomass. In the permanent plots study carried out in the upper Rio Negro region, total aboveground biomass increased from the first year to the fifth year of fallow from 708 to 3,386 g/m² (Uhl 1987). In northern Thailand, Fukushima et al. (2008) carried out a chronosequence study of fallows with stand ages from 20 to 49 years of age. Allometric relations were used to calculate the aboveground biomass. For fallow stands between 20 and 29 years of age, the values of aboveground biomass ranged from 94 to 146 t/ha, with a median of 130 t/ha. For fallows that were 30–49 years old, the aboveground biomass values were

127–305 t/ha, with a median of 236 t/ha, higher than in fallows between 20 and 29 years old.

Read and Lawrence (2003) conducted a chronosequence study conducted in the dry tropical forest of Yucatan, Mexico, sampling a total of 36 fallow stands ranging from 2 to 25 years in age in three different regions (El Refugio, Nicolas Bravo and Arroyo Negro). They reported a significant increase in total aboveground biomass with regards to fallow age: total aboveground biomass increased threefold from about 20 Mg/ha (in the first 2 years of fallow) to around 70 Mg/ha (between 12 and 25 years of fallow). Using a regression analysis, the authors suggested that the biomass of fallow stands is correlated with fallow age based on the following relationship:

$$\text{TAG biomass (Mg/ha)} = 11.431 + 2.6615^*\text{age} \left(R^2 = 0.57, \ n = 28, \ P < 0.0001\right)$$

The authors assumed in this calculation that the rate of increase of biomass would lessen when the succession processes continued, although they were unsure when the decline in growth rate would commence.

2.2.2   Total Aboveground Biomass Increases at First, but then Remains Constant

Uhl (1987) pointed out that biomass accumulation should not be linear with respect to time. The equation used to explain biomass accumulation:

$$\text{biomass} \left(g/m^2\right) = 842^*\text{age of stand in years} - 89 \left(n = 27, \ r^2 = 0.68\right)$$

which he generated from studies of the Rio Negro region in Venezuela and other tropical areas, predicted that the total aboveground biomass of fallows should reach similar levels as those found in old-growth forests in about 40 years. However, he pointed out that different results in the same region were found in a study by Saldarriaga et al. (1988). Saldarriaga et al. found that when fallows reached 40 years, the aboveground biomass could only attain about half of the levels found in old-growth forests. Uhl concluded that after 10 years of fallow, the biomass accumulation rate would decline and the liner relationship would disappear (Uhl 1987).

Other studies came to similar conclusions. A study in a forest of Sarawak, Malaysia, examined 171 fallow plots selected in 44 areas following shifting cultivation. In this study, biomass was found to increase rapidly before 10 years of fallow, although no net growth of biomass at the beginning of succession could be found (Fig. 2.1a; Jepsen 2006). Despite this, the annual accumulation rate increased from 6.6 Mg/ha in the first 2 years, to 19.5 Mg/ha by the fourth year (Fig. 2.1a). After this rapid increase, the annual accumulation rate decreased to zero when fallows aged past 10 years. Tran et al. (2010a, b) also reported a similar trend of biomass accumulation in the forest of north-western Vietnam. Referring to

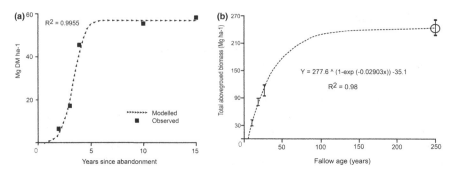

Fig. 2.1 Biomass accumulation dynamics. *Sources* **a** Jepsen (2006), **b** Tran et al. (2010b)

Fig. 2.1b, total aboveground biomass increased from 4.1 Mg/ha in 5-year-old fallows to 108.8 Mg/ha in the oldest age class (26 years). However, a slow increase was seen in the first 7 years; when fallows reached 10 years of age, the annual biomass increment rate increased rapidly to 7.77 Mg/ha. When succession proceeded, a slower growth rate was observed at first as it decreased to 5.75 Mg/ha when the fallows reached 18 years of age. The annual biomass increment rate declined further to 3.39 Mg/ha in the oldest fallow stand measured (26 years old). Tran et al. (2010b) pointed out that after 60 years of fallow, 60 % of the total aboveground biomass found in the old-growth forest was recovered. Saldarriaga et al. (1988) reported a linear increase in aboveground biomass for the first 40 years of fallow in the forest of the upper Rio Negro, but found that levels remained more or less the same for the next 40 years, ranging from 43.9 to 159 t/ha. Subsequently, levels continued to increase and ranged from 116 to 178 t/ha in 60–80-year-old fallows (Table 2.2). The stagnation of biomass accumulation might be because the growth of aboveground biomass is offset by the mortality of long-lived pioneer species (Saldarriaga et al. 1988). Uhl (1987) reached a similar conclusion: the biomass growth rate decreases at later stages due to replacement by slow-growing primary forest species. However, Uhl (1987) concluded that biomass growth rates would decrease after 10 years, earlier than that found by Saldarriaga et al. (1988).

2.3 Trees and Herbs Abundance

Delang (2007) examined the number of useful taxa found in forest fallows (1, 3, 5, 7, 9, 11 and 20 years old) of northern Laos. Delang found that the number of useable plants increased as fallows aged from 3 to 11 years, but then decreased when fallows reached 20 years of age (Fig. 2.2). El-Sheikh (2005), in a field site in Egypt, reported a rapid increase in the woody layers when fallows reach 5–6 years of age, but a slower increase in later stages of fallow (25 years of age). In eastern Madagascar, Klanderud et al. (2010) found that a positive relationship existed between the number of adult trees and tree saplings and the number of years the

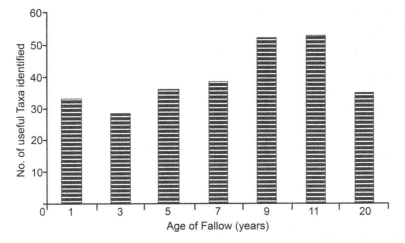

Fig. 2.2 Estimated number of usable taxa in fields with an 11-year swidden cycle. *Source* Delang (2007)

fields had been abandoned using Generalized Linear Model multiple regressions, but did not find great differences between fallows when using the Tukey's HSD post hoc test (Fig. 2.3). On the other hand, in the tropical moist forest of the upper Rio Negro Region, Saldarriaga et al. (1988) found a decreasing trend in the total number of trees with respect to fallow age, which ranged from 543 to 978 stems in 9- to 20-year-old fallows, and from 253 to 550 stems in 30- to 80-year-old fallows.

Significant differences between fallow ages can also be observed in the abundance of tree seedlings, calculated by the percentage cover of the forest floor, which show a decline with fallow duration (Fig. 2.4; Klanderud et al. 2010). The number of shrubs is also negatively correlated with years since abandonment, as estimated using Generalized Linear Model multiple regressions (Fig. 2.5; Klanderud et al. 2010).

Concerning the abundance of herbs, despite one study by Klanderud et al. (2010) in Eastern Madagascar demonstrating no great difference between fallows of different age classes, El-Sheikh (2005) found an increase in the amount of herbs in early fallow stages (1–3 years old) in the dry forests of Egypt. This trend was followed by a slight decrease from year 5 to 6, and a continued drop through later stages. Another chronosequence study conducted in the eastern lowlands of Bolivia also showed a decrease in understory vegetation when fallows aged, wherein understory cover was 100 % in the first year of fallow, but decreased to 50 % by the eighth year (Kennard 2002).

Although El-Sheikh (2005) and Klanderud et al. (2010) found an increase in the number of trees when fallows aged, a study by Saldarriaga et al. (1988) reported the opposite. However, their decreasing trend might be due to the abundance of small trees in the upper Rio Negro region (Saldarriaga et al. 1988). The authors pointed out that even though there is a gradual increase in bigger trees when fallows age, the number is not able to offset the decreasing trend of the smaller

Fig. 2.3 Species abundance of adults in fallows of different age classes, secondary, and primary forest in the Vohimana forest of Madagascar (measured in number of stems). Error bars are standard error. Bars not sharing a letter differ significantly (Tukey's hsd, P < 0.05). *Source* Klanderud et al. (2010)

Fig. 2.4 Species abundance of tree seedlings in fallows of different age classes, secondary, and primary forest in the Vohimana forest, Madagascar (measured in % cover). Error bars are standard error. Bars not sharing a letter differ significantly (Tukey's hsd, P < 0.05). *Source* Klanderud et al. (2010)

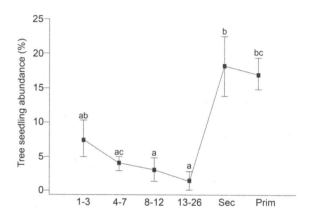

Fig. 2.5 Species abundance of shrubs in fallows of different age classes, secondary, and primary forest in the Vohimana forest of Madagascar (measured in number of stems). Error bars are standard error. Bars not sharing a letter differ significantly (Tukey's hsd, P < 0.05). *Source* Klanderud et al. (2010)

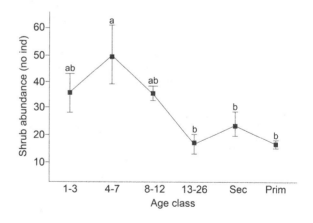

trees, as they still had high representation, even in the older fallow stages. As a result, it is possible that the unique characteristics of forests in different geographic regions could result in different growth patterns during succession.

2.4 Total Basal Area

Different researchers have reported a general trend of an increase in total basal area as fallows age. For instance, 28 fallow sites with ages ranging from one to 36 years were sampled in a semi deciduous forest of Bolivia (Toledo and Salick 2006). In the overstory, total basal area increased from 2.9 m^2/ha in 1- to 5-year-old fallows to 12.2 m^2/ha in the oldest fallows (22–36 years of age), which contained more than 50 % of the biomass measured in old-growth mature forests (21.6 m^2/ha; Table 2.2). Another chronosequence study of 36 fallow sites carried out in the tropical dry forest of the southern Yucatan Peninsular region in Mexico concluded that total basal area increased from about 10 m^2/ha in 2- to 5-year-old fallows to about 30 m^2/ha in 12- to 25-year-old fallows (Read and Lawrence 2003; Table 2.2). However, this increase was mainly driven by two dry study sites sampled—El Refugio and Nicolas Bravo. Regression analyses on the wettest site of Arroyo Negro showed no significant difference between basal area and age class of fallows (Read and Lawrence 2003).

In Colombia, another study on a tropical dry forest region also found a positive relationship between the total basal area and the fallow age (Ruiz et al. 2005). Total basal area was found to increase from 17.72 m^2/ha in the youngest fallows (less than 6 years old) to 42.05 m^2/ha in the older fallows (from 32 to 56 years of age) (Table 2.2; Ruiz et al. 2005). The fallows of that age bracket still had only less than 50 % of the basal area of oldest age class (forests older than 56 years). A chronosequence study by Kammesheidt (1998) examined 2- to 15-year-old fallows following swidden agriculture in the tropical moist forest of eastern Paraguay, and found that a positive relationship exists between basal area and fallow age. The basal area was found to increase from 3.2 m^2/ha in 2- to 5-year-old stands to 9.1 m^2/ha in 10 and 15-year-old ones (Table 2.2). The 15-year-old fallows had a basal area which was less than 50 % of that of old-growth forests (found to be 24.7 m^2/ha; Kammesheidt 1998).

Peña-Claros (2003) found in his study of tropical moist forest fallows in Bolivia that the total basal area increased from 12.3 m^2/ha in the 2-year-old fallow to 36.3 m^2/ha in the 40-year-old fallow. When fallows further regenerated to 25 years of age, 70 % of the basal area found in the old-growth forest was reached (Fig. 2.6; Peña-Claros 2003). A similar pattern was reported in a study of the tropical dry forest of the eastern lowlands of Bolivia (Kennard 2002). Kennard reported that total basal area increased with fallow duration and reached 75 % of that of the old-growth forest levels by the 23rd year of fallow (Fig. 2.6). Lebrija-Trejos et al. (2008) surveyed the tropical dry forest of southern Mexico, and showed a continuous increase in basal area throughout the succession process (ranging from 0 to 25 m^2/ha; Fig. 2.6). In this study, the increase in basal area was slower than that reported by Kennard (2002), and fallows required a longer time frame within which to reach the basal area of the mature forest. After 40 years, from 60 to 89 % of the basal area of mature forest was attained in southern Mexico (Lebrija-Trejos et al. (2008)). A longer chronosequence study of 20–100 years was

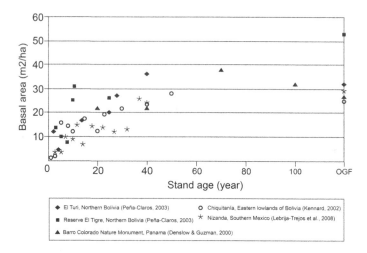

Fig. 2.6 Changes in basal area during succession in El Tigre and El Turi, northern Bolivia, Barro Colorado Nature Monument, Panama, Nizanda, southern Mexico and Chiquitanía, and the eastern lowlands of Bolivia. All individuals with ≥1 cm DBH were surveyed in northern Bolivia; all individuals with ≥5 cm DBH were surveyed in Panama; all individuals with ≥1 cm DBH, <1 cm DBH, and height ≥30 cm were surveyed in southern Mexico; all individuals with height >2 m were surveyed in the eastern lowlands of Bolivia. *Source* Figure derived from Peña-Claros (2003), Denslow and Guzman (2000), Lebrija-Trejos et al. (2008) and Kennard (2002). OGF refers to old-growth forest

conducted in the tropical moist forest of Panama. Denslow and Guzman (2000) selected sites that had been abandoned for 20, 40, 70 and 100 years, each with different land use history ranging from shifting cultivation to pastures. These authors reported that total basal area was the lowest in the youngest fallows, and continued to increase starting from year 40 until it peaked at year 70 (Fig. 2.6). At its peak, year 70 fallows had basal area values greater than those measured in the old-growth forest. Basal areas were found to subsequently drop (Fig. 2.6; (Denslow and Guzman 2000).

In a study on the upper Rio Negro, Saldarriaga et al. (1988) reported that 60- to 80-year-old fallows only had 60 % of the basal area of old-growth mature forest, a smaller number than that found by Peña-Claros (2003), Kennard (2002) or Ruiz et al. (2005). Saldarriaga et al. (1988) predicted that the time for basal areas of fallows to reach levels comparable to those of old-growth forest would be 199 years, based on the logarithmic regression:

$$\ln Y = 1.75 + 0.34 \ln X \left(n = 19, \ r^2 = 0.65, \ P < 0.0001, CV = 6.46\,\%\right),$$

where Y is the mean basal area (m²/ha), and X is the age (years). The study of permanent plots conducted in the rain forest of the upper Rio Negro region also indicated an increase of the basal area from 0.5 m²/ha in the first year to 7.0 m²/ha in the fifth year of fallow (Table 2.2; Uhl 1987).

2.5 Canopy Height

Canopy or tree height are often measured by researchers when examining the succession process in fallow fields. Different trends of change from various studies are presented below.

2.5.1 Canopy Height Increases with Fallow Age

Canopy height was measured in both dry and moist forests of Bolivia. In the dry forests of Bolivia, Kennard (2002) found that the maximum canopy height increased from about 6 m in 3-year-old fallows to about 25 m in the oldest fallows (50 years old; Fig. 2.7), and the canopy height in 23- to 40-year-old fallows attained approximately 75 % of that of the old-growth forest. On the other hand, in a Bolivian moist forest, the maximum canopy height increased from about 10 m in 10-year-old fallows to around 30 m in the oldest, 40-year-old fallows (Fig. 2.7; Peña-Claros 2003).

In Colombia and Mexico, a positive relationship also existed between mean tree height and fallow length. In Colombia, mean tree height increased from 5.8 m in fallows younger than 6 years old, to 7.7 m on 35- to 56-year-old fallows (the mean tree height of the old-growth forest was recorded to be 7.1 m) (Table 2.2; Ruiz et al. 2005). In southern Mexico's tropical dry forest, Lebrija-Trejos et al. (2008) concluded that the canopy height increased rapidly in the first 15 years, but remained stable afterwards, as shown in Fig. 2.8. After 13 years, canopy height

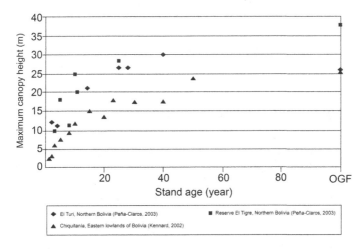

Fig. 2.7 Changes in maximum canopy height during succession in El Turi and Reserve El Tigre, northern Bolivia and Chiquitanía, and in the eastern lowlands of Bolivia. *Source* Figure derived from Peña-Claros (2003) and Kennard (2002). OGF refers to old-growth forest

Fig. 2.8 Canopy height for fallows ranging from a recently-abandoned field to a 40-yr-old field. Symbols represent actual stand values. *Lines* are time trends described by the best-fit HOF model (M I–V; *** P < 0.001). Mature forest values (M; chronosequence stand and mean with 95 % CI) are depicted for reference, but were not included in the regressions. *Source* Lebrija-Trejos et al. (2008)

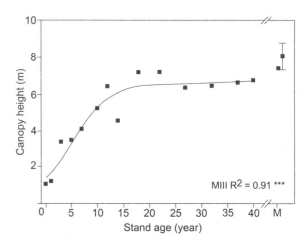

reached approximately 75 % of that found in the mature forest, and also displayed a faster growth rate than that recorded in the dry forest of Bolivia (Kennard 2002).

Other researchers in Mexico reported an increase in tree height with respect to fallow age for larger trees (dbh 5–9.9 cm and dbh larger than 10 cm) from the youngest fallows (from 2 to 5 years old) to the oldest fallows sampled (from 12 to 25 years old; Read and Lawrence 2003). However, a Tukey post hoc test showed no significant difference between stand ages and the height of smaller trees (Fig. 2.9; Read and Lawrence 2003). In the moist evergreen forest of southwestern Nigeria, tree height increased from 1.3 m in the first-year-old fallow to 5.8 m in 10-year-old fallows. These latter fallows had recovered about half of the tree height measured in old-growth forest samples (10.4 m) (Table 2.2; Aweto 1981). Nyerges (1989) conducted a chronosequence study and sampled age classes of 1, 3, 7, 8 15 17 20 25 27, 30 and over 30 years old in the tropical deciduous forest of Sierra Leone, where sites were farmed for 1 year. Canopy height was found to have a positive relationship with fallow age: it increased from 6 m in the third year to from 15 to 20 m in the 30th year of fallow (Table 2.2; Nyerges 1989).

2.5.2 No Significant Difference with Fallow Age

A chronosequence study examining the tropical dry forest of Mexico found no significant relationship between fallow age and fallow duration, basal area, or tree height (Williams-Linera et al. 2011). These authors examined five fallow sites with ages of 7 months, 8 months, 3 years, 4 years and 6 years with different land use history ranging from agriculture to pasture. Referring to Figs. 11A and C, the results of this study reported no significant difference in basal area or tree height within fallows of different age classes (from 7 months to 6 years) during early succession periods. However, the authors did report a significant difference between the overstory of old-growth forest and fallow age based on the Tukey's

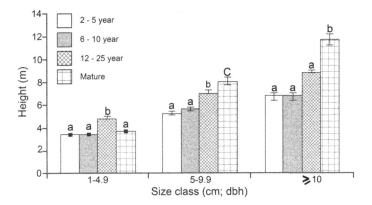

Fig. 2.9 Mean height of measured trees (n = 1395) by dbh size class Bars are means ± 1 SE. Results are of ANOVAs (all P < 0.0001) within each size class, comparing tree heights as a function of forest age. Within each size class, bars with different letters are significantly different at a level of $\alpha = 0.05$ (Tukey post hoc test). *Source* Read and Lawrence 2003

honestly significant difference (hsd; Williams-Linear et al. 2011). Differences in basal area between fallows and old-growth forest in the understory, however, showed no significant difference and almost reached the same level of the old-growth forest (Fig. 2.10a; Williams-Linear et al. 2011).

2.6 Plant Density

Studies show great variation in the changes of plant density along the succession trajectory. Some research found that plant density increases as the fallows ages, or increases and then stabilizes. Other studies have found that stem density increases initially, but then declines in early or intermediate stages, that it varies among age classes, or that it decreases with fallow age. We now review studies that found each of these contrasting results in turn.

2.6.1 Stem Density Increases, or Increases and then Stabilizes

A chronosequence study carried out in north-western Vietnam reported an increase in the stem density of tree stratum from 60 to 960 stems/ha from year 7 to the oldest (26 years old) fallows (Table 2.4; Tran et al. 2010b). Kammesheidt (1998) found a positive relationship between the mean stem density in the tree stratum (dbh $\geq$ 5 cm) and fallow age. The density was measured as 29.8/500 m^2 between 2 and 5 years, and then increased to 34.4/500 m^2 in 10- and 15-year-old fallows. This latter measurement corresponds to the stem density recorded in mature forests (33.9/500 m^2). However, the opposite was found to occur in the sapling stratum,

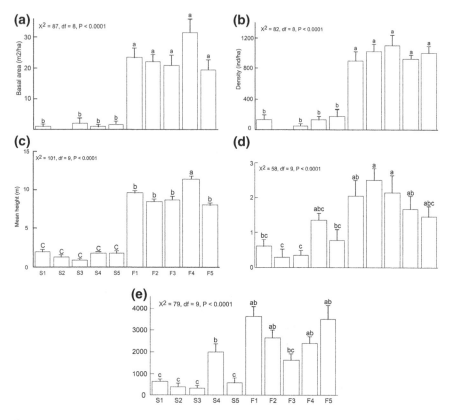

Fig. 2.10 Vegetation structure of early successional (S1-5) and forest sites (F1-5) in the tropical dry forest region of central Veracruz, Mexico. **a** Basal area, **b** density and **c** mean height of the overstory (trees ≥ 5 cm dbh). **d** Basal area and **e** density of the understory (woody plants < 5 cm dbh). Scale of the vertical axes differs between **a** and **d**, and between **b** and **e**. Different *letters* above the bars indicate significant differences between sites (P < 0.05) *Source* Williams-Linear et al. 2011

where mean stem density decreased from 68.0/100 m^2 in the young fallows to 40.3/100 m^2 in the oldest ones (Table 2.4; Kammesheidt 1998).

On the other hand, Aweto (1981) found in the moist evergreen forest fallows of Nigeria that tree density increased rapidly in the first 7 years, but then stabilized afterwards. According to Aweto's results, tree density increased from 56 stems/ha in the first year of fallow to 2,270 stems/ha in year seven, already reaching levels equivalent to those of the old-growth mature forest (2,260 stems/ha). However, these values then stabilized by 10 years to 2,670 stems/ha (Table 2.4; Aweto 1981). A similar pattern was found in the tropical dry forest of southern Mexico. Lebrija-Trejos et al. (2008) reported that stem density increased rapidly in the first 15 years of succession, but subsequently stabilized (Fig. 2.11). These authors concluded that between eight and 13 years, stem density was able to reach the

Fig. 2.11 Density of individuals for fallows ranging from a recently abandoned field to 40-year-old plots. Symbols represent actual stand values. Lines are time trends described by the best-fit HOF model (M I–V; ** P ≤ 0.01). Mature forest values (M; chronosequence stand and mean with 95 % CI) are depicted for reference but were not included in the regression calculations. *Source* Lebrija-Trejos et al. (2008)

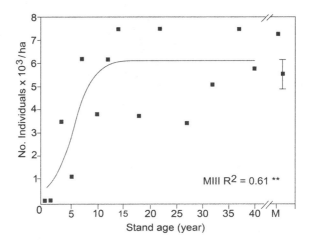

same value of the mature forest (approximately 5,511 individuals/ha). In contrast, a longer time to reach old-growth forest levels was required in forests of Nigeria (Aweto 1981).

2.6.2 *Stem Density Increases, but Shows a Decline in Early or Intermediate Stages*

Other studies also report an increase in stem density, but these studies observe a decline after reaching a peak in early to intermediate successional stages (Toledo and Salick 2006; Ruiz et al. 2005; Uhl 1987). In Bolivia, for example, plant density in the overstory increased from 55.4/0.1 ha in the early fallows (from 1 to 5 years old), peaked at 105.4/0.1 ha in the intermediate stage of 6- to 10-year-old plots, and then declined after 10 years (Table 2.4; Toledo and Salick 2006). Tree density in the overstory also demonstrated a similar picture as it increased from 52.4 stems/30 m^2 in the youngest age class (from 1 to 5 years old) to a peak of 97 stems/30 m^2 in the intermediate stage of six to 10 years (Toledo and Salick 2006). From 10 years onwards, tree density in this site declined to 69 stems/30 m^2 in the oldest age class (22–26 years old) (Toledo and Salick 2006). Similar trends were observed in the tropical dry forest of Colombia and the tropical rain forest of the upper Rio Negro. In Colombia, mean tree density of the youngest age class (under 6 years old) was about 30 stems/0.01 ha (Ruiz et al. 2005). Density increased and peaked at between 11 and 16 years of age, when mean tree density reached approximately 40 stems/0.01 ha, but then decreased after 16 years (Ruiz et al. 2005). However, in this study, there was a slight increase in the oldest age class (32–56 years old) (Fig. 2 in Ruiz et al. 2005). In the upper Rio Negro region, mean tree density (number of stems/m^2) steadily increased from the first 4 years of fallow, but then decreased after reaching the fifth year (Fig. 2.12; Uhl 1987).

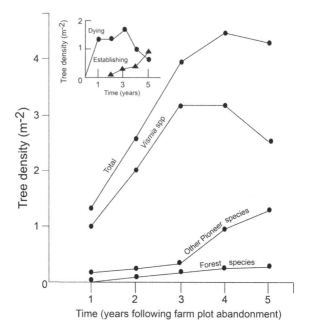

Fig. 2.12 Changes in density establishment and mortality for trees (≥2 m tall) Tree density in the 0.15 ha main study plot during the first 5 years of succession following slash-and-burn agriculture near San Carlos de Rio Negro, Venezuela. *Source* Uhl (1987)

2.6.3 Stem Density Varies Among Age Classes

Great variation in stem density among age classes was reported in a number of studies. For instance, in the tropical dry and moist forests of Bolivia, both forest types demonstrate fluctuations in the total stem density among stand ages (Kennard 2002; Peña-Claros 2003). In the eastern lowlands of Bolivia, Kennard (2002) found that total stem density (number of stems/ha) varies greatly up until year 30 of fallow. In 23-year-old stands, stem density was surprisingly high. The author suggested that the high values in year 23 might have been due to variations in field conditions, such as differences in soil types, location, fire occurrence and land history. From year 30 total stem density gradually dropped (Fig. 2.14). Similar to results reported by Kennard (2002), Peña-Claros (2003) reported great variation among age classes, with no clear correlation found between stem density (number of stems/ha *1000) and fallow age in either Reserve El Tigre or in El Turi of Bolivia (Fig. 2.13). On the other hand, a chronosequence study conducted in the dry tropical forest in Mexico using Tukey's honestly significant difference test showed that woody plant density had similar values among age classes from 7 months to 6 years old in both overstory and understory, although a great difference could be seen between fallows and old-growth forests in the overstory (Fig. 2.10b, e; Williams-Linera et al. 2011). However, differences between fallows

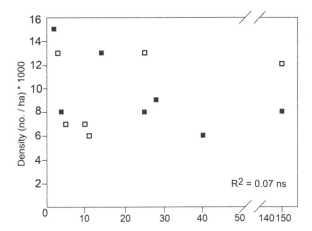

Fig. 2.13 Density of all individuals ≥1 m height. *Open circles* are data from the Reserve El Tigre and *filled circles* are data from El Turi. *Source* Peña-Claros (2003)

and old-growth forest were found to be smaller in the understory, where some plots had already reached values comparable with those of old-growth forests (Williams-Linera et al. 2011).

2.6.4 Stem Density Decreases with Fallow Age

Stem density was found to decline in the study conducted in the southern Yucatan Peninsular region of Mexico, despite the positive relationship between total aboveground biomass and fallow age (Read and Lawrence 2003). For 2- to 5-year-old plots, stem density ranged from 13,000 to 33,000 stems/ha, whereas in the old-growth forest, average stem density was found to be 12,900 stems/ha. Read and Lawrence (2003) reported a greater decline in density with fallow age in Nicolas Bravo and Arroyo Negro, while the driest region of El Refugio had a smaller drop compared to the other two sites (Fig. 2.14).

In a chronosequence study conducted in north-western Vietnam, Tran et al. (2011) examined the density of seedlings, saplings and trees of four species of trees (*Wendlandia paniculata*, *Schima wallichii*, *Camellia tsaii* and *Lithocarpus ducampii*) on fallows of different age. The researchers found that the seedling density of the first three species first increased slightly, and then decreased sharply, while the seedlings of *L. ducampii* remained more or less constant (Fig. 2.15). A similar pattern was observed with the saplings for the first three species, with the sapling density increasing from 5-year old fallows to 10-year old fallows, and then decreasing (no saplings of *L. ducampii* were found). On the other hand, tree density decreased from 10-year old fallows onward (again, no *L. ducampii* were found). According to the authors, the decrease in seedling and sapling density

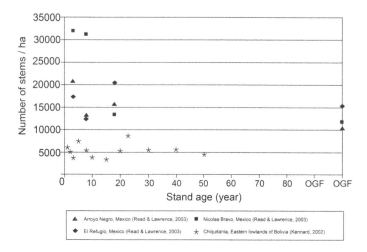

Fig. 2.14 Changes in stem density during succession in El Refugio, Arroyo Negro and Nicolas Bravo, Mexico and Chiquitanía, and the eastern lowlands of Bolivia. Trees, palms, and lianas >1 cm DBH were surveyed in Mexico; individuals with height>2 m were surveyed in Eastern lowlands of Bolivia. Figure derived from Read and Lawrence (2003) and Kennard (2002). OGF refers to old-growth forest

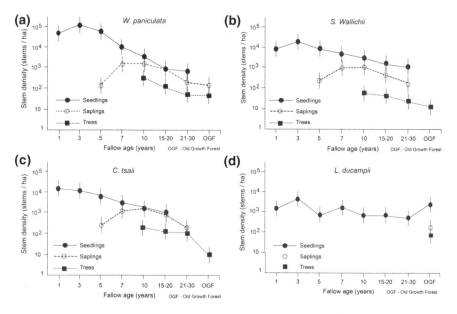

Fig. 2.15 Change in seedling, sapling and tree density (± SD) **a** *Wendlandia paniculata*, **b** *Schima wallichii*, **c** *Camellia tsaii*, and **d** *Lithocarpus ducampii* in different ages of secondary forest regrowth and in the old-growth forest in north-western Vietnam. *Vertical bars* indicate ±standard deviation. *Source* Tran et al. (2011)

might have been due to the increase in total crown area, as not sufficient sunlight penetrated through the canopy, which decreased the amount of forest-floor light, and may have inhibited the growth of seedlings. Tran et al. (2011) suggested that their results agreed with the findings of Whitmore (1983), who concluded that the decrease in the amount of light was responsible for tree mortality.

2.7 Discussion and Conclusions

Table 2.1 summarises the various patterns of change in total aboveground biomass, total basal area and canopy height in different places, and Table 2.2 presents the various results reported by the case studies we examined. Referring to Table 2.2, a general trend in the increase in total aboveground biomass as fallows age is observed, in spite of some researchers suggesting that this increase will eventually reach a threshold. The time needed to reach this threshold also varies from study to study (Table 2.1). For example, in Malaysia, total aboveground biomass on fallows became constant after 10 years, whereas in the upper Rio Negro region of Colombia and Venezuela, total aboveground biomass only became constant after 40 years.

Different studies also reported variations in the time needed for aboveground biomass on fallows to reach levels comparable to those of old-growth forests. For instance, in forests of Puerto Rico, the fallows failed to reach levels comparable to those of old growth forests, even after 80 years (Brown and Lugo 1990b). Likewise, in the upper Rio Negro region, 80-year-old fallows only had about two thirds the amount of total aboveground biomass of the old-growth forest, with a mean of only 179 t/ha for 80-year-old stands, while total aboveground biomass of old-growth forest reached 255 t/ha. Thus, the authors inferred that it would take a projected 189 years in order for total aboveground biomass in fallows to attain values equivalent to those of old-growth forests (Saldarriaga et al. 1988). Similarly, studies of forests of Mexico and northern Thailand concluded that the aboveground biomass of fallows could not reach equivalent values to those of old-growth forests even within 50 years (Hughes et al. 1999; Fukushima et al. 2008). Read and Lawrence (2003) studied the dry tropical forests of Yucatan in Mexico, and concluded that, assuming biomass and fallow age have a linear relationship, total aboveground biomass would recover to the same level of old-growth forest in 65–120 years.

In other cases, researchers found that total aboveground biomass does recover almost to the same levels as old-growth forests within 30–40 years. For example in northern Thailand, old-growth forest above-ground biomass measured 185–260 t/ha, with a median value of 254 t/ha, while 30- to 49-year-old fallows only measured 127–305 t/ha, with a median of 236 t/ha, only slightly less than that of the old-growth forest (Table 2.2; Fukushima et al. 2008). Fallows in the forest of Central Burma were found to be the fastest to reach the equivalent amount of

Table 2.1 Summary of changes in total aboveground biomass, total basal area, tree/canopy height when fallows age and the number of years needed to attain old-growth forest levels

Place	Forest Type	Changes in total aboveground biomass along succession	Changes in total basal area along succession	Changes in tree/canopy height along succession	Years to reach comparable amount of aboveground biomass with OGF	Years to reach comparable amount of total basal area with OGF	Years to reach comparable amount of tree/canopy height with OGF	Source
Central Burma	Mixed deciduous forest	Increase	/	/	About 40 years	/	/	Fukushima et al. (2007)
Upper Rio Negro region of Venezuela	Tropical moist forest	Increase	Increase	/	/	/	/	Uhl (1987)
Northern Thailand	Lower montane forest / Primary evergreen forest	Increase	/	/	More than 49 years	/	/	Fukushima et al. (2008)
Mexico	Tropical dry forest	Increase	Increase	/	65–120 years	/	/	Read and Lawrence (2003)
Malaysia	/	Increase but remained constant after 10 years	/	/	/	/	/	Jepsen (2006)
Upper Rio Negro region of Colombia and Venezuela	Tropical moist forest	Linear increase but remained constant after 40 years	Increase	/	189 years	199 years; Reaching 60% after 60–80 years	/	Saldarriaga et al. (1988)
North-western Vietnam	Evergreen broad-leaved forest	Slow increase in first 7 years; rapid increase from 10 year and growth rate reduce from 18 years onwards	/	/	More than 60 years	/	/	Tran et al. (2010a, b)

(continued)

Table 2.1 (continued)

Place	Forest Type	Changes in total aboveground biomass along succession	Changes in total basal area along succession	Changes in tree/canopy height along succession	Years to reach comparable amount of aboveground biomass with OGF	Years to reach comparable amount of total basal area with OGF	Years to reach comparable amount of tree/canopy height with OGF	Source
Bolivia	Semi-deciduous forest	/	Increase	/	/	More than 36 years	/	Toledo and Salick (2006)
Colombia	Tropical dry forest	/	Increase	Increase	/	More than 56 years	More than 56 years	Ruiz et al. (2005)
Bolivia	Tropical moist forest	/	Increase	/	/	Reaching 70% after 25 years	/	Peña-Claros (2003)
Eastern lowland Bolivia	Tropical dry forest	/	Increase	Increase	/	Reaching 75% after 23 years	Reaching 75% after 23–40 years old	Kennard (2002)
Southern Mexico	Tropical dry forest	/	Continuous increase	Increase in the first 15 years then remains stable	/	Reaching 60–89% after 40 years	Reaching 75% after 13 years	Lebrija-Trejos et al. (2008)
Panama	Tropical moist forest	/	Increase	/	/	Before 70 years	/	Denslow and Guzman (2000)
Paraguay	Tropical moist forest	/	Increase	/	/	/	/	Kammesheidt (1998)
Nigeria	Moist evergreen forest	/	/	Increase	/	/	/	Aweto (1981)
Sierra Leone	Tropical deciduous forest	/	/	Increase	/	/	/	Nyerges (1989)
Mexico	Tropical dry forest	/	No significant difference	No significant difference	/	/	/	Williams-LInera et al. (2011)

Results of different studies

aboveground biomass as the old-growth forest (Table 2.1). For example, Fukushima et al. (2007) reported that the total aboveground biomass on fallows already contained 34.8 % of that of the old-growth forest in the fifth year, and reached 93.6 % in year 40. The authors suggested that this rapid increase in the early stages might be due to the presence of bamboo, which can survive fire exposure. Similar results were reported in other studies carried out in tropical forests, where authors concluded that 30- to 40-year-old fallows could reach aboveground biomass levels comparable to those of old-growth forests (Snedaker 1970; Lugo et al. 1974; Scott 1977). As shown in Table 2.1, fallows in the tropical dry forests of Mexico (Read and Lawrence 2003) had a faster growth rate than those in the tropical moist forest of the upper Rio Negro region (Saldarriag et al. 1988). However, due to a lack of information, it is not clear if tropical dry forests have a faster aboveground biomass growth rate than humid forests. In short, the examples we highlight demonstrate a gradual increase in total aboveground biomass with fallow length, but different rates of biomass accumulation have been reported by different studies. Other factors such as climate and land use patterns in the past (Brown and Lugo 1990a; Silver et al. 2000, Guariguata and Ostertag 2001), soil fertility and soil condition (Johnson et al. 2000; Moran et al. 2000; Zarin et al. 2001; Chazdon 2003) likely have an impact on the biomass accumulation rate. We will address this in later sections of the book.

According to Table 2.2, in general, total basal area was found to increase during succession. However, Fig. 2.6 showed that in Panama, total basal area dropped when fallows reached 100 years of age. Additionally, some studies demonstrated that the old-growth forests do not necessarily have a greater basal area than fallows (Fig. 2.6). As a result, some variations in the successional pathway could be observed among studies. On the other hand, the time needed to attain levels similar to the old-growth forest differed among studies. For instance, in the tropical dry forests of Colombia, eastern lowlands of Bolivia and southern Mexico, growth rates of total basal area varied greatly. Of the three sites, the eastern lowlands of Bolivia required the shortest time to reach old-growth forest levels (23 years; Table 2.1). Differences could not only be observed among dry forests, but also among moist forests. The tropical moist forest of Bolivia had the fastest growth rate of total basal area among moist forests, with 25 years to reach 70 % of the old-growth forest value (Table 2.1). In contrast, in the moist forest of the upper Rio Negro region of Colombia and Venezuela, 60–80 years was required.

As mentioned earlier, the trends of changes in total basal area and changes in total aboveground biomass are thought to be related (Chazdon et al. 2007; Clark and Clarl 2000). Uhl (1987), Read and Lawrence (2003) and Saldarriaga et al. (1988) measured both aboveground biomass and basal area in their studies (Table 2.1). Results of Uhl (1987) and Read and Lawrence (2003) seem to agree in the sense that they both report aboveground biomass and basal area as increasing along the succession. On the other hand, Saldarriaga et al. (1988) found a trend in total basal area, but did not find a constant increase in total aboveground biomass (levels remained constant after 40 years). Saldarriaga et al. (1988) examined the longest chronosequence of 80 years; in contrast, Uhl (1987) and Read and Lawrence (2003) did not consider fallows beyond 25 years of age (Table 2.2). As

Table 2.2 Values of total aboveground biomass, total aboveground biomass accumulation rate, total basal area and canopy/tree height from different studies

Forest type	Soil	Age (yr)	Total aboveground biomass (t/ha)	Total aboveground biomass accumulation rate (t/ha/yr)	Total basal area (m²/ha)	Canopy/ tree height (m)	Source
Mixed deciduous forest	Ultisols	1	0.3	/	/	/	Fukushima et al. (2007)[a]
		2	0.6	/	/	/	
		5	53.9	/	/	/	
		10	67.8	/	/	/	
		15	79.0	/	/	/	
		18	82.9	/	/	/	
		>40	147.6	/	/	/	
		OGF	156.6	/	/	/	
Tropical rain forest	Oxisols/ Ultisols	1	708[b]	5–9	0.5	/	Uhl (1987)
		2	1282[b]		1.8	/	
		3	1995[b]		4.0	/	
		4	2861[b]		5.7	/	
		5	3386[b]		7.0	/	
Lower montane forest / Primary evergreen forest	Coarse sandy loamy soil	20–29	94–146 (median 130)	/	/	/	Fukushima et al. (2008)
		20–29		/	/	/	
		20–29		/	/	/	
		20–29		/	/	/	
		20–29		/	/	/	
		30–39	127–305 (median 236)	/	/	/	
		30–39		/	/	/	
		30–39		/	/	/	
		40–49		/	/	/	
		40–49		/	/	/	
		40–49		/	/	/	
		OGF	185–260 (median 254)	/	/	/	
		OGF		/	/	/	
		OGF		/	/	/	
		OGF		/	/	/	
		OGF		/	/	/	
		OGF		/	/	/	
		OGF		/	/	/	
Moist evergreen forest	Ferrallitic tropical soils	1	/	/	/	1.3	Aweto (1981)
		3	/	/	/	2.5	
		7	/	/	/	4.7	
		10	/	/	/	5.8	
		OGF	/	/	/	10.4	
Tropical deciduous forest	/	1	/	/	/	/	Nyerges (1989)
		3	/	/	/	6	
		7	/	/	/	8	
		8	/	/	/	/	
		15	/	/	/	10	
		17	/	/	/		
		20	/	/	/	13	
		25	/	/	/		
		27	/	/	/		
		30	/	/	/	15–20	
		>30	/	/	/	/	

(continued)

Table 2.2 (continued)

Forest type	Soil	Age (yr)	Total aboveground biomass (t/ha)	Total aboveground biomass accumulation rate (t/ha/yr)	Total basal area (m²/ha)	Canopy/ tree height (m)	Source
Tropical dry forest	/	2[e]	22.36[c]	/	9.3	/	Read and
		3	10.60[c]	/	4.4	/	Lawrence
		4	16.30[c]	/	7.7	/	(2003)
		5	17.48[c]	/	7.2	/	
		8	29.07[c]	/	10.5	/	
		8	20.53[c]	/	9.1	/	
		10	39.67[c]	/	15.9	/	
		12	45.56[c]	/	20.7	/	
		12	54.70[c]	/	18.3	/	
		12	60.98[c]	/	24.6	/	
		OGF	120.34[c]	/	36.5	/	
		OGF	138.54[c]	/	37.7	/	
		OGF	123.68[c]	/	37.7	/	
		3[f]	18.17[c]	/	17.5	/	
		5	34.44[c]	/	14.6	/	
		5	64.06[c]	/	28.0	/	
		6	20.74[c]	/	9.2	/	
		8	24.87[c]	/	10.8	/	
		16	50.02[c]	/	19.2	/	
		18	20.41[c]	/	20.6	/	
		24	97.57[c]	/	34.3	/	
		25	94.20[c]	/	26.5	/	
		25	60.46[c]	/	29.2	/	
		OGF	163.30[c]	/	43.8	/	
		OGF	133.11[c]	/	39.1	/	
		OGF	157.88[c]	/	45.0	/	
		4[g]	12.34[c]	/	5.9	/	
		5	31.60[c]	/	13.8	/	
		5	25.17[c]	/	12.8	/	
		7	11.89[c]	/	5.1	/	
		8	62.55[c]	/	18.8	/	
		9	20.42[c]	/	10.2	/	
		15	36.19[c]	/	24.4	/	
		18	27.59[c]	/	64.2	/	
		OGF	136.31[c]	/	38.0	/	
		OGF	118.16[c]	/	30.9	/	

(continued)

Table 2.2 (continued)

Forest type	Soil	Age (yr)	Total aboveground biomass (t/ha)	Total aboveground biomass accumulation rate (t/ha/yr)	Total basal area (m²/ha)	Canopy/tree height (m)	Source
Between northern Amazonian forests and southern dry forests	/	2	/	6.6[d]	/	/	Jepsen (2006)
		4	/	19.5[d]	/	/	
		10	/	0.0[d]	/	/	
Tropical moist forest	Oxisols, Ultisols Poor in nutrient	9	43.9	/	11.17	/	Saldarriaga et al. (1988)
		11	52.9	/	11.52	/	
		12	81.8	/	17.31	/	
		14	53.4	/	11.12	/	
		20	83.3	/	17.50	/	
		20	61.5	/	14.45	/	
		20	63.8	/	17.06	/	
		20	97.6	/	18.70	/	
		30	53.8	/	11.70	/	
		35	109	/	19.61	/	
		35	108	/	20.07	/	
		40	159	/	22.98	/	
		60	116	/	17.68	/	
		60	197	/	31.04	/	
		60	138	/	24.74	/	
		80	134	/	22.31	/	
		80	178	/	26.44	/	
		80	144	/	23.92	/	
		80	142	/	23.20	/	
		OGF	223	/	30.45	/	
		OGF	262	/	35.62	/	
		OGF	264	/	36.21	/	
		OGF	271	/	36.95	/	
	Oxisols	2	/	/	3.2	/	Kammesheidt (1998)
		3	/	/		/	
		4	/	/		/	
		5	/	/		/	
		10	/	/	9.1	/	
		15	/	/		/	
		OGF	/	/	24.7	/	
Evergreen brad-leaved forest	Ferralic Acrilsols Acidic, poor in nutrient	1	/	/	/	/	Tran et al. (2010b)
		3	/	/	/	/	
		5	4.1[c]	0.82[d]	/	/	
		7	12.4[c]	4.15[d]	/	/	
		10	35.7[c]	7.77[d]	/	/	
		18	81.7[c]	5.75[d]	/	/	
		26	108.8[c]	3.39[d]	/	/	
		OGF	240.5[c]	/	/	/	

(continued)

Table 2.2 (continued)

Forest type	Soil	Age (yr)	Total aboveground biomass (t/ha)	Total aboveground biomass accumulation rate (t/ha/yr)	Total basal area (m²/ha)	Canopy/tree height (m)	Source
Semideciduous forest	/	1–5	/	/	2.9	/	
		6–10	/	/	8.9	/	
		12–20	/	/	12.7	/	
		22–36	/	/	12.2	/	
		OGF	/	/	21.6	/	
Tropical dry forest	/	<6	/	/	17.72	5.8	Ruiz et al.
		6–10	/	/	24.88	6.3	(2005)
		11–16	/	/	20.46	6.5	
		17–31	/	/	26.37	6.6	
		32–56	/	/	42.05	7.7	
		>56	/	/	111.09	9.0	

OGF refers to old-growth forest
[a] excluding herbs and grass
[b] g/m²
[c] Mg/ha
[d] Mg/ha/yr
[e] Region: El Refugio
[f] Region: Nicolas Bravo
[g] Region: Arroyo Negro
[h] Overstory total basal area

suggested by Uhl (1987), biomass growth rate should not be linear in the long run; thus, the different trends of biomass and basal area observed in the upper Rio Negro region (Saldarriaga et al. 1988) could be due to the longer chronosequence. Because of these uncertainties, it would be useful to carry out additional research on longer periods of succession.

Table 2.1 and Fig. 2.7 showed the changes in tree and canopy heights during succession that are reported in different studies. A positive relationship between tree and canopy height and age along succession was observed in most studies, except that of Williams-Linear et al. (2011) in the tropical dry forests of Mexico. The Mexico study showed no great difference among fallows. This might have been due to the short period of fallows observed by the authors (only 1- to 5-year-old fallows), as tree heights might not have increased significantly during early stages of fallow, as shown in Table 2.2. Aweto (1981), Nyerges (1989) and Ruiz et al. (2005) examined longer periods of succession in their studies, and reported greater differences in tree height between early and late fallows. More and longer studies in the dry forests of Mexico will improve our understanding of the forest regeneration pattern along the successional process in this region, and help us determine if it is congruent with patterns found in other forests.

The time needed to reach a height similar to that found in old-growth forests varies among studies (Table 2.1). In southern Mexico and the eastern lowlands of Bolivia (both sites are in tropical dry forest) different times were needed to reach

Table 2.3 Summary of changes in plant density when fallows aged, and the number of years needed to reach the level of old-growth forest from different studies

Place	Forest type	Changes in stem density along succession	Years to reach comparable amount with old-growth forest	Source
North-western Vietnam	Evergreen broad-leaved forest	Increase in overstory stem density	Reaching more than 90 % after 26 years	Tran et al. (2010a, b)
Paraguay	Tropical moist forest	Increase in tree stratum; Decrease in sapling stratum	10–15 years in tree stratum	Kammesheidt (1998)
Bolivia	Semi-deciduous forest	Increase but decline after 10 years in overstory	About 5 years	Toledo and Salick (2006)
Colombia	Tropical dry forest	Increase before 11–16 years	Less than 6 years	Ruiz et al. (2005)
Upper Rio Negro region of Venezuela	Tropical moist forest	Increase but decline after 4 years	/	Uhl (1987)
Nigeria	Moist evergreen forest	Increase in first 7 years then remains stable	7 years	Aweto (1981)
Eastern lowland Bolivia	Tropical dry forest	Great variation	/	Kennard (2002)
Bolivia	Tropical moist forest	Great variation, no correlation found	/	Peña-Claros (2003)
Mexico	Tropical dry forest	No significant difference in overstory and understory	/	Williams-Linera et al. (2011)
Mexico	Tropical dry forest	Decrease	/	Read and Lawrence (2003)
Southern Mexico	Tropical dry forest	Increase in first 15 years then remains stable	8–13 years	Lebrija-Trejos et al. (2008)
North-western Vietnam	Evergreen broad-leaved forest	Decrease	/	Tran et al. (2011)[a]

[a] Four species were examined: *W. paniculata*, *S. wallichii*, *C. tsaii* and *L. ducampii*

75 % of the height of old-growth forest (13 years in southern Mexico and from 23 to 40 years in the eastern lowlands of Bolivia). There are only a few studies that provided data on these regions (see Table 2.1), thus, it is difficult to draw conclusions about the forest regeneration rates.

Table 2.4 Values of stem densities of overstory and understory, and total stem density from different studies

Forest type	Soil	Age (yr)	Stem density/ha (overstory)	Stem density/ha (understory)	Total stem density/ha	Source
Tropical moist forest	Oxisols	2	68.0[c]	29.8[d]	/	Kammesheidt (1998)
		3			/	
		4			/	
		5			/	
		10	40.3[c]	34.4[d]	/	
		15			/	
		OGF	30.8[c]	33.9[d]	/	
Evergreen brad-leaved forest	Ferralic Acrilsols Acidic and poor in nutrient	1	/	/	/	Tran et al. (2010b)
		3	/	/	/	
		5	/	/	/	
		7	60	/	/	
		10	730	/	/	
		18	830	/	/	
		26	960	/	/	
		OGF	1020	/	/	
Between northern Amazonian forests and southern dry forests	/	1–5	55.4[a]	239.4[b]	/	Toledo and Salick (2006)
		6–10	105.4[a]	292.0[b]	/	
		12–20	82.7[a]	314.7[b]	/	
		22–36	75.9[a]	232.4[b]	/	
		OGF	69.5[a]	222.4[b]	/	
Moist evergreen forest	Ferrallitic tropical	1	/	/	56	Aweto (1981)
		3	/	/	512	
		7	/	/	2270	
		10	/	/	2670	
		OGF	/	/	2260	

[a] ind./0.1 ha
[b] ind./30 m^2
[c] Sapling stratum 1–4.9 cm dbh/100 m^2
[d] Tree stratum, dbh $\geq$5 cm/500 m^2

The different changes reported in levels of plant density are summarized in Table 2.3. Many researchers observed that younger fallows have higher tree density, lower canopy height and lower basal area (Saldarriaga et al. 1988; Denslow and Guzman 2000; Guariguata and Ostertag 2001). Chazdon et al. (2007) also pointed out that the stem density of fallows is often higher than that of old-growth mature forests. Studies of the eastern lowlands of Bolivia (Kennard 2002), lowland Costa Rica (Guariguata et al. 1997) and Nigeria (Aweto 1981) agreed with this observation. However, others observed variation along the chronosequence, and it is clear that change in stem density along succession varies greatly among the studies reviewed here. Ruiz et al. (2005) suggested that stem density is likely to peak at the intermediate age class, which is demonstrated in their study in the dry forests of Colombia, and a study by Kennard (2002) in the eastern lowlands of Bolivia. Table 2.4 reports the values of plant densities from various studies, and also shows patterns of variations along the chronosequence, and peaks at the intermediate age classes, for both overstory and understory composition. As suggested by Chazdon et al. (2007), stem density does not show a predictable pattern. Thus, we can conclude that in the successional process, fluctuation in tree density level is greater than both basal area and aboveground biomass. The great variation of stem density shows that a wide range of factors influence its change. These factors are likely to be soil, land use intensity and other locational or temporal factors not addressed in the papers reviewed (Chazdon et al. 2007; Kennard 2002).

A few studies estimated similar lengths in the number of years needed for fallows to reach comparable plant densities as old-growth forests. Fallows in the tropical dry forest of Colombia, semi-deciduous forests of Bolivia and moist evergreen forests of Nigeria had similar plant density growth rates, requiring from 5 to 7 years to reach levels of the old growth forest (Table 2.3). In the tropical dry forests of southern Mexico, however, between 8 and 13 years were required. As shown in Table 2.3, the longest period of time to attain old-growth forest plant density was reported for the evergreen broad-leaved forest of north-western Vietnam. Based on these results, we are unable to conclude which type of forest demonstrates a faster rate of plant density regeneration.

This chapter presented a bigger picture of the changes in forest structure during secondary forest succession following shifting cultivation. Comparisons on forest structure among regions and forest types are rarely found in the literature. By reviewing the results from different researches, we can conclude that total above ground biomass, total basal area, canopy height, and plant density do not show a consistent development pathway, which implies that the conditions or characteristics of the forest plays a part in altering the regeneration of forest structure during succession. Chazdon et al. (2007) also highlighted the importance of local species composition on influencing the changes in vegetation, which is seldom mentioned or examined in studies of forest succession. If local or regional characteristics were taken into consideration, in addition to the differences in fallow age, we would be able to better understand the patterns of forest structural changes during secondary succession.

References

Aweto AO (1981) Secondary succession and soil fertility restoration in south-western Nigeria I, Succession. J Ecol 69:601–607
Brown S (1997) Estimating biomass and biomass change of tropical forests: A primer. FAO, Rome
Brown S, Lugo AE (1990a) Tropical secondary forests. J Trop Ecol 6:1–32

Brown S, Lugo AE (1990b) Effects of forest clearing and succession on the carbon and nitrogen content of soils in Puerto Rico. Plant Soil 124:53–64

Chazdon RL (2003) Tropical forest recovery: legacies of human impact and natural disturbances. Perspect Plant Ecol Evol Syst 6:51–71

Chazdon RL, Letcher SG, van Breugel M, Martínez-Ramos M, Bongers F, Finegan B (2007) Rates of change in tree communities of secondary neotropical forests following major disturbances. Philos Trans R Soc B 362:273–289

Clark DB, Clark DA (2000) Landscape-scale variation in forest structure and biomass in a torpical rain forest. For Ecol Manag 137:185–198

Delang CO (2007) Ecological succession of usable plants in an eleven-year fallow cycle in Northern Lao P. D. R. Ethnobotany Res Appl 5:331–350

Denslow JS, Guzman GS (2000) Variation in stand structure, light, and seedling abundance across a tropical moist forest chronosequence, Panama. J Veg Sci 11:201–212

El-Sheikh MA (2005) Plant succession on abandoned fields after 25 years of shifting cultivation in Assuit, Egypt. J Arid Environ 61:461–481

Fukushima M, Kanzaki M, Hla MT, Minn Y (2007) Recovery process of fallow vegetation in the traditional Karen swidden cultivation system in the Bago Mountain Range, Myanmar. Southeast Asian Stud 45(3):317–333

Fukushima M, Kanzaki M, Hara M, Ohkubo T, Preechapanya P, Choocharoen C (2008) Secondary forest succession after the cessation of swidden cultivation in the montane forest area in Northern Thailand. For Ecol Manag 255:1994–2006

Guariguata MR, Ostertag R (2001) Neotropical secondary forest succession: changes in structural and functional characteristics. For Ecol Manag 148:185–206

Guariguata MR, Chazdon RL, Denslow JS, Dupuy JM, Anderson L (1997) Structure and floristics of secondary and old-growth forest stands in lowland Costa Rica. Plant Ecol 132:107–120

Hughes RF, Kauffman JB, Jaramillo VJ (1999) Biomass, carbon, and nutrient dynamics of secondary forests in a humid tropical region of Mexico. Ecology 80(6):1892–1907

Jepsen MR (2006) Above-ground carbon stocks in tropical fallows, Sarawak, Malaysia. For Ecol Manag 225:287–295

Johnson CM, Zarin DJ, Johnson AH (2000) Post-disturbance aboveground biomass accumulation in global secondary forests. Ecology 81:1395–1401

Kammesheidt L (1998) The role of tree sprouts in the restoration of stand structure and species diversity in tropical moist forest after slash-and-burn agriculture in Eastern Paraguay. Plant Ecol 139(2):155–165

Kennard DK (2002) Secondary forest succession in a tropical dry forest: patterns of development across a 50-year chronosequence in lowland Bolivia. J Trop Ecol 18:53–66

Klanderud K, Mbolatiana HAH, Vololomboahangy MN, Radimbison MA, Roger E, Totland Ø, Rajeriarison C (2010) Recovery of plant species richness and composition after slash-and-burn agriculture in a tropical rainforest in Madagascar. Biodivers Conserv 19:187–204

Lebrija-Trejos E, Bongers F, Pérez-García EA, Meave J (2008) Successional change and resilience of a very dry tropical deciduous forest following shifting agriculture. Biotropica 40:422–431

Lugo AE, Brinson M, Ceranevivas M, Gist C, Inger R, Jordan C, Lieth H, Milstead W, Murphy P, Smythe N, Snedaker S (1974) Tropical ecosystem structure and function. In: Farnworth EG, Golley FB (eds) Fragile ecosystems. Springer, New York, pp 67–111

Moran EF, Brondizio E, Tucker JM, da Silva-Fosberg MC, McCracken S, Falesi I (2000) Effects of soil fertility and land-use on forest succession in Amazonia. For Ecol Manag 139:93–108

Nyerges AE (1989) Coppice swidden fallows in tropical deciduous forest: biological, technological, and sociocultural determinants of secondary forest successions. Hum Ecol 17(4):379–400

Ogawa H, Yoda K, Ogino K, Kira T (1965) Comparative ecological studies on three main types of forest vegetation in Thailand. Nature and Life in SE Asia. Fauna and Flora Research Society, Kyoto

Peña-Claros M (2003) Changes in forest structure and species composition during secondary forest succession in the Bolivian Amazon. Biotropica 35(4):450–461

Read L, Lawrence D (2003) Recovery of biomass following shifting cultivation in dry tropical forests of the Yucatan. Ecol Appl 13(1):85–97

Ruiz J, Fandiño MC, Chazdon RL (2005) Vegetation structure, composition, and species richness across a 56-year chronosequence of dry tropical forest on Providencia Island, Colombia. Biotropica 37(4):520–530

Saldarriaga JG, West DC, Tharp ML, Uhl C (1988) Long-term chronosequence of forest succession in the upper Rio Negro of Colombia and Venezuela. J Ecol 76:938–958

Scott G (1977) The importance of old-field succession biomass increments to shifting cultivation. Gt Plains Rocky Mt Geogr J 6:318–327

Silver WL, Ostertag R, Lugo AE (2000) The potential for carbon sequestration through reforestation of abandoned tropical agricultural and pasture lands. Restor Ecol 8:394–407

Snedaker SC (1970) Ecological studies on tropical moist forest succession in eastern lowland Guatemala. Ph.D. thesis, University of Florida, Gainesville

Toledo M, Salick J (2006) Secondary succession and indigenous management in semideciduous forest fallows of the Amazon Basin. Biotropica 38(2):161–170

Tran P, Marincioni F, Shaw R (2010a) Catastrophic flood and forest cover change in the Huong river basin, central Viet Nam: A gap between common perceptions and facts. J Environ Manag 91(11):2186–2200

Tran VD, Osawa A, Nguyen TT (2010b) Recovery process of a mountain forest after shifting cultivation in Northwestern Vietnam. For Ecol Manage 259:1650–1659

Tran VD, Osawa A, Nguyen TT, Nguyen BV, Bui TH, Cam QK, Le TT, Diep XT (2011) Population changes of early successional forest species after shifting cultivation in Northwestern Vietnam. New Forest 41:247–262

Uhl, C. 1987. Factors controlling succession following slash-and-burn agriculture in Amazonia. J Ecol 75:377–407

Whitmore TC (1983) Secondary succession from seed in tropical rain forests. For Abs 44:767–779

Williams-Linera G, Alvarez-Aquino C, Hernández-Ascención E, María T (2011) Early successional sites and the recovery of vegetation structure and tree species of the tropical dry forest in Veracruz, Mexico. New Forests 42:131–148

Zarin DJ, Ducey MJ, Tucker JM, Salas WA (2001) Potential biomass accumulation in Amazonian regrowth forests. Ecosystems 4:658–668

Chapter 3
Species Richness and Diversity

Abstract This chapter reviews the literature on species richness and species diversity. The two concepts are closely related, but are not synonyms. Species richness is estimated dividing the number of species by the geographic area. On the other hand, species diversity is a function of the number of species present (i.e. species richness or number of species) and the evenness with which the individuals are distributed among these species (i.e. species evenness, species equitability, or abundance of each species). In this chapter, we divide the literature under the subheading 'total species richness', 'tree and herb species richness', and 'total species diversity'. The review of the literature reveals that some studies observed an increase in species richness and species diversity when fallows age, while others found an increase at the beginning but then a drop in levels (or constant levels) in later stages. A few studies even concluded that there is no difference among age classes. Therefore, we are not able to extrapolate a general model for how total species richness is impacted by fallow duration. Similarly, the time needed for total species richness to recover to levels found in old-growth forest also shows great variations.

Keywords Species richness · Species diversity · Intermediate disturbance hypothesis · Successional stages · Shannon-wiener index · Simpson's index

3.1 Introduction

Studies of ecological succession of secondary forests often look at the vegetation of fallows in terms of species richness and species diversity. The two concepts are closely related, but are not synonyms. The term "species richness" was first introduced by McIntosh (1967). Sanjit and Bhatt (2005) pointed out that species

C. O. Delang and W. M. Li, *Ecological Succession on Fallowed Shifting Cultivation Fields*, SpringerBriefs in Ecology, DOI: 10.1007/978-94-007-5821-6_3, © The Author(s) 2013

richness can be expressed numerically, through an account of the number of species (Hurlbert 1971). Species richness is estimated dividing the number of species by the geographical area, a measure developed and used extensively by Simpson (1964), among others. On the other hand, according to Sanjit and Bhatt (2005, p. 557), species diversity is "a function of the number of species present (i.e. species richness or number of species) and the evenness with which the individuals are distributed among these species (i.e. species evenness, species equitability, or abundance of each species)" (Margalef 1958; Loyd and Ghelardi 1964; Pielou 1966; Spellerberg 1991). Species richness is easier to measure than species diversity, as measuring species diversity implies accounting for both 'species richness' and 'species evenness, species abundance or species equitability' (Sanjit and Bhatt 2005). Because of these differences between these two terms, they have to be used carefully to avoid misinterpretation (Sanjit and Bhatt (2005) pointed out that these two terms are often used interchangeably and without a clear definition in most of the literature, which created confusion).

According to the intermediate disturbance hypothesis, the highest diversity should be found in regions with an intermediate level of disturbance (Connell 1978). Connell (1978) explained that diversity would increase when the time interval away from disturbance increased, as more species could colonize and reach maturity. However, species diversity would decrease when the land is left for further regeneration because of the replacement of some species by other, more efficient ones (Connell 1978). In addition, if the competitiveness between species were the same, those with higher resistance to physical alterations of the land would be more likely to survive in disturbed habitats. As a result, diversity should drop in the long-run only if other factors do not interfere to maintain it (Connell 1978). Under these assumptions, it is suggested that species richness peaks at the intermediate successional stage, and that the oldest stage (old-growth forest) will be less diverse (Ruiz et al. 2005; Sheil 2001; Sheil and Burslem 2003). In the following sections, we review studies that examine species richness and species diversity: first total species richness, second tree and herb species richness, and third species diversity.

3.2 Total Species Richness

There does not seem to be conclusive results as to the relationship between total species richness and fallow age. Some studies found that species richness increases with fallow age. Other studies found that species richness increases at first, then drops off or stays constant, and other again found no significant difference in species richness as fallows age. We review each of these cases in turn.

3.2.1 Species Richness Increases with Fallow Age

In the rain forest of Nigeria, Ross (1954) selected 3 sample plots of fallows following shifting cultivation, aged 5, 14 and 17 years since last abandonment. The total number of species in the sample plots was shown to increase with fallow age, with 60 in the 5 year-old fallow, 63 in the 14th year, and 67 in the oldest, 17-year-old fallow (Table 3.2). Another chronosequence study conducted in forests of Vietnam also showed a positive relationship between total species richness and fallow length (Tran et al. 2010b, 2011). Tran et al. (2010a, 2011) analyzed 51 sites with 1–30 years fallow after shifting cultivation in the evergreen broad-leaved forest of north-western Vietnam. The authors concluded that the total number of species increased from four on the 1-year-old fallows to 18 on the 10-year-old stands, and reached 35 at the oldest fallow stage (Table 3.2). The 26-year-old fallows, had 49 % of the total number of species found in the old-growth forest (Tran et al. 2010b). The authors found a nearly linear relationship between total number of species and fallow age, despite a gradual reduction in the rate of increase in species richness as the secondary forest ages. Tran et al. (2010a) suggested that about 60 years are needed for fallows to reach a level of species richness comparable with that of old-growth forest. This result agrees with a review by Brown and Lugo (1990a), which concluded that for fallow fields in tropical regions, less than 80 years would be needed to equal the amount of species found in the old-growth forest.

Another chronosequence study carried out in the dry tropical forest of Colombia also found a linear increase in species richness as the fallow aged (Ruiz et al. 2005). The number of species increased from 23 in the youngest (<6-year-old fallows) to 40 in the oldest fallows (32–56 years old) (Table 3.2). In that study, 17–31-year-old fallows contained 75 % of the number of species found in old-growth forest, while 32–56-year-old fallows contained 90 % of the number of species found in old-growth forests (Ruiz et al. 2005). At the same time, the mean species density increased from the youngest fallows to the 56-year-old fallows (Fig. 3.1). Species density and richness was also measured in the study of dry tropical forest of southern Mexico, where Lebrija-Trejos et al. (2008) examined 15 fallow sites that cropped for 1 or 2 years, with a chronosequence of 1–40 years. Their results showed that the number of species gradually increases with fallow age, increasing from 15 to 58 species (Fig. 3.2). After 40 years, species density was found to be 24.6 species/100 m^2, reaching 80–96 % of that in mature forest. This recovery was much faster than that recovered in forests in Vietnam (Tran et al. 2010b) (see Table 3.1).

Another chronosequence study conducted in the lowland tropical rainforest and semi-deciduous forest of Mexico looked at eight sites with 1- to 5-year-old fallows after corn cultivation with one year cropping (Breugel et al. 2007). In the first 5 years of fallow, species richness appeared to have a positive relationship with fallow length—the number of species increases from 24 to 54 as the fallows ages

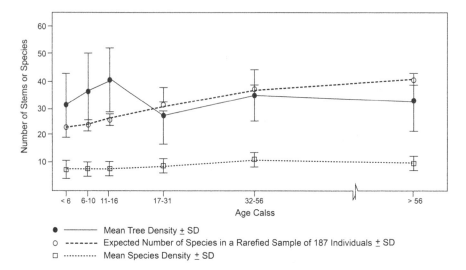

Fig. 3.1 Mean species density, mean tree density and expected number of species. Mean species density (number of species/0.01 ha), mean tree density (number of individuals/0.01 ha), and expected number of species in a rarefied sample of 187 individuals of woody species > 2.5 cm DBH, on Providencia, specific by age class. The midpoint for each age class was plotted on the linear X-axis. For the group >56 year, 80 year was assumed. *Source* Ruiz et al. (2005)

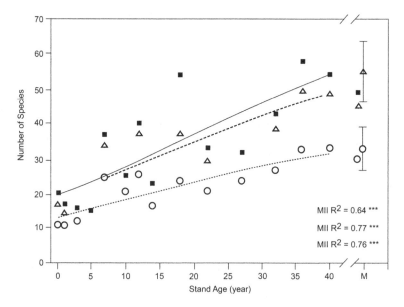

Fig. 3.2 Species density and species richness. Species density (*closed squares, continuous line*) and species richness rarefied to 175 individuals (*close triangles, medium-dash line*) and 80 individuals (*open circles, short-dash line*) for fallows ranging from recently abandoned field to 40-years old. Symbols represent actual stand values. Lines are time trends described by the best-fit HOF model (M I–V; *** $P < 0.001$). *Source* Lebrija-Trejos et al. (2008)

Table 3.1 Summary of changes in total species richness when fallows aged, and number of years needed to reach old-growth forest levels from different studies

Place	Forest type	Changes in total species richness along succession	Years to reach comparable amount with old-growth forest	Source
North-western Vietnam	Evergreen broad-leaved forest	Nearly linear increase	About 60 years	Tran et al. (2010b, 2011)
Tropics	/	Increase	<80 years	Brown and Lugo (1990b)
Upper Rio Negro region of Colombia and Venezuela	Tropical moist forest	Increase, then stay relatively constant at later stages	20–40 years	Saladarriaga et al. (1988)
Colombia	Tropical dry forest	Linear increase	More than 56 years	Ruiz et al. (2005)
Mexico	Lowland tropical rain forest and semi-deciduous forest	Increase	/	Breugel et al. (2007)
Nigeria	Moist evergreen forest	Increase but stabilize in early stage	7–10 years	Aweto (1981)
Nigeria	Tropical rain forest	Increase	/	Ross (1954)
Paraguay	Tropical moist forest	Increase	After 2–5 years in sapling stratum	Kammesheidt (1998)
South-western Madagascar	Dry deciduous forest	Peak at intermediate stage, then slightly decline	/	Raharimalala et al. (2010)
Bolivia	Tropical moist forest	Increase, then stay constant at intermediate stage	10–20 years	Peña-Claros (2003)
Egypt	Dry forest	Increase at the beginning, then decline at early stage	/	El-Sheikh (2005)
Southern Mexico	Tropical dry forest	Gradual increase	Reaching 80–96 % after 40 years	Lebrija-Trejos et al. (2008)
Northern Thailand	Lower montane forest/Primary evergreen forest	No great difference between age classes	/	Fukushima et al. (2008)
Eastern Madagascar	Tropical rainforest	No great difference between age classes, but great increase from oldest fallows to old-growth forest	More than 30 years	Klanderud et al. (2010)

Table 3.2 The number of tree, shrub, herbaceous, liana, and all species found on fallows with different age classes in different studies. OGF means to old-growth forest

Forest type	Soil	Age (yr)	Number of species					Source
			Trees	Shrubs	Herbaceous	Liana	All	
Evergreen broad-leaved forest	Ferralic Acrilsols Acidic and poor nutrient content	1	/	/	/	/	4	Tran et al. (2010b, 2011)
		3	/	/	/	/	7	
		5	/	/	/	/	12	
		7	/	/	/	/	16	
		10	/	/	/	/	18	
		18	/	/	/	/	23	
		26	/	/	/	/	35	
		OGF	/	/	/	/	72	
Tropical moist forest	Oxisols, Ultisols	9	50[c]/5[d]	/	/	/	55	Saladarriaga et al. (1988)
		11	33[c]/8[d]	/	/	/	41	
		12	43[c]/15[d]	/	/	/	58	
		14	56[c]/7[d]	/	/	/	63	
		20	65[c]/15[d]	/	/	/	80	
		20	76[c]/17[d]	/	/	/	93	
		20	50[c]/11[d]	/	/	/	61	
		20	63[c]/24[d]	/	/	/	87	
		30	77[c]/13[d]	/	/	/	90	
		35	68[c]/19[d]	/	/	/	87	
		35	70[c]/18[d]	/	/	/	88	
		40	75[c]/16[d]	/	/	/	91	
		60	62[c]/13[d]	/	/	/	75	
		60	70[c]/25[d]	/	/	/	95	
		60	66[c]/23[d]	/	/	/	89	
		80	72[c]/17[d]	/	/	/	89	

(continued)

Table 3.2 continued

Forest type	Soil	Age (yr)	Number of species					Source
			Trees	Shrubs	Herbaceous	Liana	All	
		80	73[c]/25[d]	/	/	/	98	
		80	60[c]/30[d]	/	/	/	90	
		80	79[c]/22[d]	/	/	/	101	
		OGF	75[c]/30[d]	/	/	/	105	
		OGF	66[c]/14[d]	/	/	/	80	
		OGF	96[c]/28[d]	/	/	/	124	
		OGF	63[c]/23[d]	/	/	/	86	Kammesheidt (1998)
	Oxisols	2	/	/	/	/	39[a]/19[b]	
		3						
		4						
		5						
		10					44[a]/33[b]	
		15						
		OGF					37[a]/42[b]	
Tropical dry forest	/	<6					23	
		6–10					31	
		11–16					36	
		17–31					35	
		32–56					40	
		>56					49	Ruiz et al. (2005)
Moist evergreen forest	Ferrallitic tropical	1	2.2				39.2	
		3	7				42.5	
		7	14.5				56.6	Aweto (1981a)

(continued)

Table 3.2 continued

Forest type	Soil	Age (yr)	Number of species					Source
			Trees	Shrubs	Herbaceous	Liana	All	
Tropical rain forest	/	10	19	/	/	/	53.5	Ross (1954)
		Mature secondary	24.6	/	/	/	60.6	
		5	/	/	/	/	60	
		14	/	/	/	/	63	
		17	/	/	/	/	67	
Dry deciduous forest	Sedimentary deposits	1–5	4	4	21	3	32	Raharimalala et al. (2010)
		6–10	12	5	22	6	45	
		11–20	20	13	27	9	69	
		21–30	31	13	30	12	86	
		31–40	37	11	21	12	81	
		>40	29	10	21	13	73	
Tropical dry forest	Lithosols and Haplic Phaeozems	1	/	/	/	/	15–58	Lebrija-Trejos et al. (2008)
		3	/	/	/	/		
		5	/	/	/	/		
		7	/	/	/	/		
		10	/	/	/	/		
		12	/	/	/	/		
		14	/	/	/	/		
		18	/	/	/	/		
		22	/	/	/	/		
		27	/	/	/	/		
		32	/	/	/	/		
		37	/	/	/	/		
		40	/	/	/	/		

(continued)

Table 3.2 The number of tree, shrub, herbaceous, liana, and all species found on fallows with different age classes in different studies. OGF means to old-growth forest

Forest type	Soil	Age (yr)	Number of species					Source
			Trees	Shrubs	Herbaceous	Liana	All	
Lower montane forest/Primary evergreen forest	Coarse sandy loamy soil	20–29					24–35	Fukushima et al. (2008)
		20–29						
		20–29						
		20–29						
		20–29						
		30–39					20–36	
		30–39						
		30–39						
		40–49						
		40–49						
		40–49						
Lowland tropical rain forest and semideciduous forest	/	1					24–54	Breugel et al. (2007)
		2						
		3						
		4						
		5						

[a] Sapling stratum (1-4.9 dbh)
[b] Tree stratum (dbh $\geq$ 5 cm)
[c] Trees with dbh $\geq$ 1 cm
[d] Trees with dbh $\geq$ 10 cm

from one to five year old (Breugel et al. 2007). This could be explained by the higher rate of species recruitment than mortality. Breugel et al. (2007) examined the plots after 18 months of the first field census, and found that the total number of dead trees ranged from 12 to 20 species, with a mean of 16 after 18 months, while the total number of species recruited during this 18 months period was 19 to 40, with a mean of 28. As a result, the number of newly recruited species exceeds the number of species lost, which resulted in an increase in total species richness (Breugel et al. 2007). Kammesheidt (1998) observed the changes in species richness of fallows in tropical moist forest of Paraguay by examining both sapling and tree stratum (Table 3.2). Species richness in both stratums showed a positive relationship with fallow age, which increased from 39 species in 2- to 5-year-old stands to 44 species in 10- and 15-year-old ones in the sapling stratum. Species richness reached a similar level as old-growth forest (37 species) after 10–15 years (Kammesheidt 1998). In the tree stratum, the increase in species number is more significant, as it changed from 19 species in the young fallows (2- to 5-year-old) to 33 species in the older fallows (10 and 15 years old). Unlike the sapling stratum, species richness did not attain the same level as the old-growth forest in the tree stratum, even after 15 years of fallow (Kammesheidt 1998).

Results from the tropical rainforest of Eastern Madagascar were largely congruent (Klanderud et al. 2010). Klanderud et al. used Tukey's HSD post hoc test, and reported that total species richness remained constant at around 20 species regardless of fallow age until age 10–12, but then increased in 13-26-year-old fallows, which is when the fields continue to regenerate into secondary forests. Primary forests had the highest number of species in total (Fig. 3.3). However, using a Generalized Linear Model, multiple regressions—used to measure the relative importance of different variables for species richness—revealed that total species richness had a positive relationship with the number of years since abandonment (Fig. 3.3). The authors concluded that fallow length could influences changes in species richness, but total species richness in fallows would not reach a comparable value as old-growth forest, even for the oldest fallow (26 years old) (see Fig. 3.4).

3.2.2 Species Richness Increases at First, then Drops Off or Stays Constant

The pattern of constant increase in species richness has not been observed by all researchers. Many also found a different pattern of changes for total species richness along the successional process. Unlike the examples from Vietnam (Tran et al. 2010b) and Colombia (Ruiz et al. 2005), which revealed a linear relationship, others found an increase in the number of species from early to intermediate successional stages, with a decline or stabilisation when fallows aged. This was

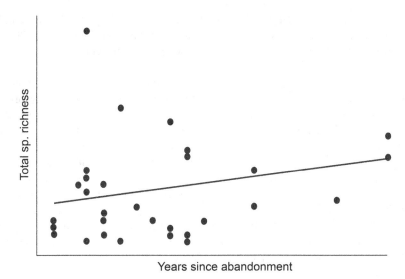

Fig. 3.3 Relationship between total species richness and significant ($P < 0.05$) fallow parameters By GLM multiple regression analysis. *Source* Klanderud et al. (2010)

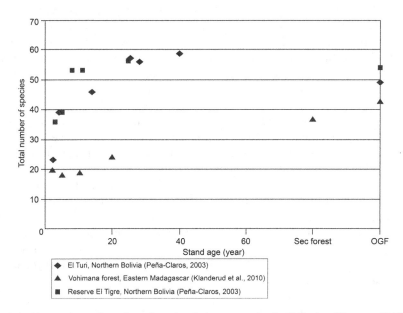

Fig. 3.4 Changes in total number of species during succession in El Turi and Reserve El Tigre, northern Bolivia, and the Vohimana forest, Eastern Madagascar. OGF refers to old-growth forest. *Source* Figure derived from Peña-Claros (2003) and Klanderud et al. (2010)

observed in the dry forest of the south-western coast of Madagascar (Raharimalala et al. 2010), the tropical moist forest of the Bolivian Amazon (Peña-Claros 2003), the dry forest of Egypt (El-Sheikh 2005) and the moist evergreen forest of Nigeria (Aweto 1981).

In Madagascar, Raharimalala et al. (2010) found that total species richness increased from 32 species in 1–5 years of fallow up to 81 species in 21–30-year-old fallows, but dropped to 73 species in 40-year-old forest as succession proceeded further (Table 3.2). In the tropical moist forest of the Bolivian Amazon, 2 to 40 year-old fallows from two different sites, El Tigre forest reserve and El Turi, were selected for a chronosequence study (Peña-Claros 2003). A total of 23 species were found in 2-year-old fallows, a number which continued to increase up to 25 years, after which it remained relatively constant. The total number of species could reach a number comparable to mature forest somewhere between 10 and 20 years (Fig. 3.4). Saladarriaga et al. (1988) concluded from their study in the tropical moist forest of the Upper Rio Negro Region in Colombia and Venezuela that the number of species increased rapidly during the first 20–30 years, and that a similar number of species compared to the old-growth forest could be attained after 20–40 years (see Table 3.2). Their study reported a slower growth rate than that in the tropical moist forest of the Bolivian Amazon. Aweto (1981) found a general increase in the total number of species with regards to fallow age, which tended to stabilize in the early successional stage. The chronosequence study was conducted in the forest of Nigeria, where a total of forty one-, three-, seven-, and 10-year-old fallows were examined. The total number of species increased from 39.2 in the first year of fallow to 56.6 in the 7-year-old ones, and remained relatively constant afterwards, with 10- year-old fallows containing 53.3 species (Table 3.2). A comparable level of total species richness to that of mature forest was attained in the 7th year, with a slightly lower number (total number of species in mature forest: 60.6) (Aweto 1981). A slightly higher increase in the rate of total species richness could be observed when compared to the forest in Bolivia (Peña-Claros 2003) (see Table 3.1).

El-Sheikh (2005) found that species richness rises in the early fallow stage from 1 to 2 years, but gradually decreases until later stages (from the 5th year of fallow on) in the dry forest of Egypt. Compared to Madagascar (Raharimalala et al. 2010) and the Bolivian Amazon (Peña-Claros 2003), the total number of species in the Egyptian forests surveyed declined much earlier than in these other forests. The author found an increase in the percentage of tree and species as fallows aged, likely due to the decrease in the effect of disturbance. In such undisturbed or climaxed environments, Margalef (1968); Bazzaz (1975) and Naveh and Whittaker (1979) all suggested that total species richness would decline. The earlier decline of total species richness might indicate that the disturbance effect is reduced in the earlier stage in Egypt.

3.2.3 No Significant Difference Between Age Classes

Other successional patterns have been found elsewhere. In the lower montane forest of northern Thailand, a chronosequence study was conducted on 11 fallows with stand ages ranging from 20 to 49 years, following 1 year of upland rice farming (Fukushima et al. 2008). The number of species found on these fallows did not differ between age classes. For age class ranging from 20 to 29-years-old, the number of species found was between 24 and 35, while for 30–49-year-old stands, the number of species ranged from 20 to 36.

3.3 Tree and Herb Species Richness

Similarly to the results for total species richness, studies found different regeneration patterns for tree species richness. In the following pages, we first review the studies that have found an increase in species richness as fallow age progresses, then we review those studies that have found that the number of tree species increases up to a certain age, and then starts to drop off.

3.3.1 Species Richness Increases with Fallow Age

In the long-term permanent plots, which 'directly [document] the rates of change through monitoring vegetation dynamics over time in particular forest stands' (Chazdon et al. 2007, pp. 274) a study conducted by Uhl (1987; Voeks 1996) in the upper Rio Negro region of the Amazon Basin examined forest tree species. This study found that forest tree species increase steadily from the first year of fallow to the 5th year (Fig. 3.5), but will need many decades for species richness to reach a level similar to old-growth forests. Despite this, the richness of pioneer tree species that are taller than or equal to 2 m did not show great differences from the first year of fallow to the 5th year. Saldarriaga et al. (1988) reached similar conclusions: the time needed for species richness of bigger tress with dbh (diameter at breast height) $\geq$ 10 cm to reach levels similar to those of old-growth forest were found to be at least 40 years in the upper Rio Negro Basin, and the number of bigger tree species ranged from 5 to 30 from the youngest (9 years old) to the oldest fallow stands (80 years old) (see Table 3.2). However, the authors also found that for smaller trees (dbh > 1 cm), only 10–20 years are needed for recovery, which is a much faster regeneration rate than that of larger trees (Saldarriaga et al. 1988). In Eastern Madagascar, Klanderud et al. (2010) found that the species richness of adult trees is positively correlated with years since abandonment by using the Generalized Linear Model multiple regressions, although no great difference could be observed by Tukey's HSD post hoc test among fallow age classes (Fig. 3.5).

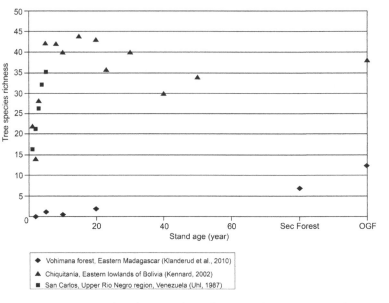

Fig. 3.5 Changes in total number of tree species during succession in Vohimana forest, Eastern Madagascar, San Carlos, Venezuela and Chiquitanía, Eastern lowlands Bolivia Trees with dbh > 10 cm were surveyed in Madagascar; trees with height ≥ 2 m were surveyed in Venezuela; trees with height > 2 m were surveyed in Bolivia. OGF refers to old-growth forest. *Source* Figure derived from Klanderud et al. (2010), Uhl (1987) and Kennard (2002)

A chronosequence study conducted in the moist evergreen forest of Nigeria also found an increase in the number of tree species when fallows age (Aweto 1981). Aweto (1981) found that from the first year of fallow, the number of tree species was found to be 2.2 stems/900 m^2. This increased to 19 stems/900 m^2 on 10 year-old-fallows, and to 24.6 stems/900 m^2 in old-growth forest (see Table 3.2).

3.3.2 Species Richness Peaks at Early to Intermediate Stages

The constant increase in tree species richness has not been observed by all of the researchers reported here. For example, 12 fallow stands following shifting-cultivation, with a chronosequence of 1–50 years, were examined in the dry forest of the Eastern lowlands of Bolivia. Tree species richness significantly increased during early succession and reached 75 % of that of the old-growth forest by the first 5 years of fallow (Kennard 2002). Yet, after 15–20 years, tree species richness started to decline (Fig. 3.5). Raharimalala et al. (2010) also found a drop in tree species richness in a study in south-western Madagascar: the number of tree species increased from four during 1- to 5-year old fallows, to 37 in 31–40-year-old fallows. However, these numbers dropped to 29 after 40 years. The number of tree species decreased at a later stage to levels lower than those recorded in the dry forest of Bolivia (see Table 3.2).

Another study carried out in the semi-deciduous forest of Bolivia reported a rapid increase in species density of the overstory from 7.0 (1–5-years-old fallows) to 26.6 (oldest fallows ranging in age from 22 to 36 years old) after studying 28 fallows with 1–5, 6–10, 12–20 and 22–36 years following cultivation (Toledo and Salick 2006). The overstory species density displayed an even higher value than that of the old-growth forest, which was 23.6 (Toledo and Salick 2006). Nevertheless, changes in species density of understory trees with regards to fallow age revealed a completely different picture. Toledo and Salick (2006) observed a slight variation within age classes ranging from 25.1 to 33.1, demonstrating that the species density of understory species does not have a strong correlation with fallow length.

Chronosequence studies in eastern Madagascar (Klanderud et al. 2010) and the upper Rio Negro region (Saldarriaga et al. 1988) found similar patterns when examining pioneer herbaceous species richness on ageing fallows. Klanderud et al. (2010) observed a peak of herbaceous species richness in the first 3 years of fallow (10 species), which then decreased gradually to seven species in the oldest fallows (Fig. 3.6). In addition, Saldarriaga et al. (1988) found that the number of pioneer plants species decreased with fallow age, and dominated the field in the first 20 years but then dropped off in later stages (see Table 3.2).

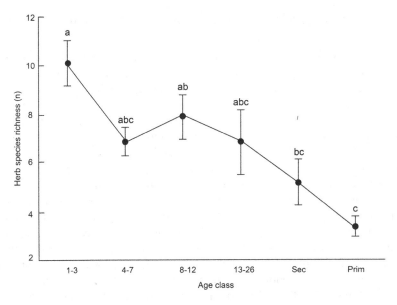

Fig. 3.6 Species richness of herbs in fallows of different age classes in secondary and primary forest plots in the Vohimana forest, Madagascar. Error bars are standard error. Bars not sharing a letter differ significantly (Tukey's hsd, P < 0.05) *Source* Klanderud et al. (2010)

3.4 Total Species Diversity

Sanjit and Bhatt (2005, p. 557) define species diversity as 'a function of the number of species present (i.e. species richness or number of species) and the evenness with which the individuals are distributed among these species (i.e. species evenness, species equitability, or abundance of each species)' (Margalef 1958; Loyd and Ghelardi 1964; Pielou 1966; Spellerberg 1991). There are two main indices to measure species diversity: the Shannon-Wiener index (Shannon and Weaver 1949), commonly known as the Shannon index, and the Simpson's index (Simpson 1949). The Shannon index depends greatly on sample size and places more weight on species richness, while the Simpson's index places more weight on species evenness (Sanjit and Bhatt 2005). In this chapter, we will discuss the different findings regarding changes in species diversity during succession, and will examine if the intermediate disturbance hypothesis (IDH) is congruent with results generated from different studies.

A number of studies have used the Shannon index as a means to measure species diversity (Tran et al. 2010b; Fukushima et al. 2007; Metzger 2003; Saldarriaga et al. 1988; Ruiz et al. 2005; Peña-Claros 2003; Uhl 1987; Voeks 1996; Fukushima et al. 2008; El-Sheikh 2005; Toledo and Salick 2006; Kammesheidt 1998; Lebrija-Trejos et al. 2008) and found various changes in total species diversity along the succession trajectory.

3.4.1 Species Diversity Increases Rapidly, then Remains Constant

In the upper Rio Negro region (Saldarriaga et al. 1988) and the Bolivian Amazon (Peña-Claros 2003), species diversity as measured by the Shannon index increases rapidly in the first 20 years of fallow, and subsequently remains relatively constant. Saladarriage et al. (1988) found that the Shannon index ranged from 3.32 in a young fallow to 5.01 in a 20-year-old plot. In older plots, it ranged from 4.98 to 5.90 (Table 3.4). Fallows in the Bolivian Amazon demonstrated an increase in species diversity with stand age, but species diversity stabilised after 20–25 years (Fig. 3.7; Peña-Claros 2003). However, these authors found variation in species diversity changes between forest layers in the Bolivian Amazon. The Shannon diversity index measure of these plots showed greater variation with time in the canopy layer than in the lower layers, and the lowest diversity was found in the young fallows (8–14-years-old; Fig. 3.8). Regression analyses of species diversity of understory did not show a correlation with fallow duration, but species diversity in both the sub-canopy and canopy layers were significantly correlated with fallow duration.

A similar trend was reported in another chronosequence study done in the semi-deciduous forest of Bolivia (Toledo and Salick 2006). Toledo and Salick found that the Shannon index increased greatly from the early successional stages to the

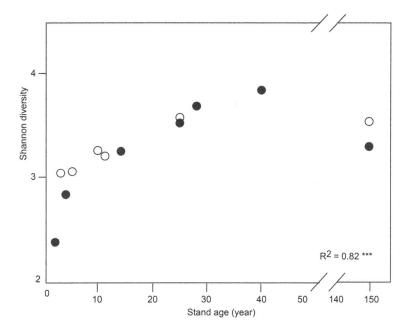

Fig. 3.7 Shannon diversity index of secondary and mature forests in northern Bolivia. *Open circles* are data from the Reserve El Tigre and *filled circles* are data from El Turi. *Source* Peña-Claros (2003)

later stage in the overstory (from 5.2 in 1–5-years-old fallows to 28.6 in 22–36 year-old fallows), where values reached those measured in old-growth forest (Shannon index in old-growth forest of overstory: 27). In contrast, understory showed a decreasing trend in species diversity, with the youngest fallows possessing the highest value (33.7), and subsequently dropping to 18.7 in the oldest fallows (22–36 years old). These latter values were comparable to those found in old-growth forest (Table 3.4).

In north-western Vietnam, Tran et al. (2010a) also found a major increase in species diversity in early stages of fallow, with the Shannon index rising from 0.96 in the first year of fallow to 2.18 in year seven. However, the growth rate started to drop when fields approached year 10, and stayed relatively constant until 26 years (Table 3.4). Compared to fallows of the upper Rio Negro (Saldarriaga et al. 1988) and the Bolivian Amazon (Peña-Claros 2003), species diversity on fallows of Vietnam (Tran et al. 2010b) slowed down more rapidly (see Table 3.4). Moreover, a study in Nigeria showed an even earlier peak of species diversity (Aweto 1981). In this study, the author used Simpson's index to examine species diversity and found that diversity increased during the first 7 years from 0.37 in the 1 year-old fallow to 0.71 in the 7-year-old fallow, when it reached a peak (the maximum level). Afterwards, species diversity remained constant, with a level of species diversity comparable to those recorded in old-growth forest (0.71 reached by 7-year-old fallows; Table 3.4).

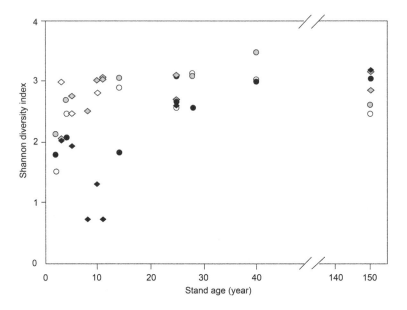

Fig. 3.8 Species diversity (based on the Shannon diversity index) for each combination of forest layer and stand age. Diamonds represent data from the Reserve El Tigre and *circles* are data from El Turi. *Open symbols* are used for the understory layer, *gray symbols* for the sub-canopy layer, and *filled symbols* for the canopy layer. *Source* Peña-Claros (2003)

3.4.2 Species Diversity Increases with Fallow Age

Other studies observed a constant increase in species diversity. A chronosequence study conducted in tropical dry forest of Colombia found a constant rate of increase in the Shannon index from 2.31 in fallows under 6 years of age, to 2.5 in 11–16-year-old fallows, to 3.19 in the oldest fallows (32–56 years old). These oldest fallows were able to reach similar values as those of the old-growth forest (3.35; Table 3.4; Ruiz et al. 2005). Another study carried out in the tropical dry forest of southern Mexico found a gradual increase in the Shannon diversity index with regards to fallow age. Index measurements ranged from 1.36 to 3.34 along the 40 year chronosequence (Table 3.4; Lebrija-Trejos et al. 2008). According to the authors, when fallows reached 40 years, the Shannon index shows that biodiversity is 85–90 % of that of undisturbed forests, a time frame which was similar to that in the dry forest of Colombia (Table 3.3; Ruiz et al. 2005).

In a long-term permanent plot study carried out in the upper Rio Negro region, the author examined species diversity of the tropical moist forest following slash-and-burn agriculture, using both the Shannon index and Simpson's diversity index

Table 3.3 Summary of changes in total species diversity when fallows aged and the number of years needed to reach the level of old-growth forest from different studies

Place	Forest type	Changes in total species diversity along succession	Years to reach a comparable number than old-growth forest	Source
Bolivia	Tropical moist forest	Increase but growth rate reduce after 20–25 years	10–30 years	Peña-Claros (2003)
Upper Rio Negro region of Colombia and Venezuela	Tropical moist forest	Increase but growth rate reduce after 20 years	About 80 years	Saldarriaga et al. (1988)
Bolivia	Semi-deciduous forest	Increase in overstory; Decrease in understory	22–36 years	Toledo and Salick (2006)
North-western Vietnam	Evergreen broad-leaved forest	Increase but growth rate reduce after 10 years	About 60 years	Tran et al. (2010)
Nigeria	Moist evergreen forest	Increase than remains constant after 7 years	7 years	Aweto (1981)
Colombia	Tropical dry forest	Increase	About 56 years	Ruiz et al. (2005)
Upper Rio Negro region of Venezuela	Tropical moist forest	Increase	/	Uhl (1987)
Egypt	Dry forest	Increase in the beginning but decrease after 3 years	/	El-sheikh (2005)
Southern Mexico	Tropical dry forest	Gradual increase	Reaching 86–90 % after 40 years	Lebrija-Trejos et al. (2008)
Northern Thailand	Lower montane forest/Primary evergreen forest	Decrease	20–29 years	Fukushima et al. (2008)

Table 3.4 Values of species diversity index measures from different studies

Forest type	Soil	Age (yr)	Shannon diversity index	Simpson's index	Source
Tropical moist forest	Oxisols, Ultisols	9	3.86	0.89	Saladarriaga et al. (1988)
	Poor in nutrient	11	3.32	0.85	
		12	4.10	0.90	
		14	4.63	0.94	
		20	4.84	0.94	
		20	5.01	0.94	
		20	4.49	0.93	
		20	4.38	0.89	
		30	5.18	0.96	
		35	4.98	0.94	
		35	5.30	0.96	
		40	5.38	0.96	
		60	5.25	0.96	
		60	4.89	0.94	
		60	5.09	0.995	
		80	5.18	0.95	
		80	4.94	0.96	
		80	5.90	0.94	
		80	5.02	0.97	
		OGF	5.30	0.96	
		OGF	5.45	0.95	
		OGF	4.75	0.97	
		OGF	5.25	0.95	

(continued)

Table 3.4 (continued)

Forest type	Soil	Age (yr)	Shannon diversity index	Simpson's index	Source
	Oxisols	2	1.79[d]/2.14[e]	/	Kammesheidt (1998)
		3		/	
		4		/	
		5		/	
		10	2.33[d]/2.84[e]	/	
		15		/	
Semideciduous forest	/	OGF	2.34[d]/3.09[e]	/	Toledo and Salick (2006)
		1–5	5.2[a]/33.7[b]	/	
		6–10	13.3[a]/15[b]	/	
		12–20	20.2[a]/10.9[b]	/	
		22–36	28.6[a]/18.7[b]	/	
		OGF	27[a]/19.3[b]	/	
Evergreen broad-leaved forest	Ferralic Acrilsols Acidic, poor in nutrient	1	0.96	/	Tran et al. (2010)
		3	1.09	/	
		5	1.47	/	
		7	2.18	/	
		10	2.21	/	
		18	2.72	/	
		26	3.09	/	
		OGF	3.44	/	
Moist evergreen forest	Ferrallitic tropical	1	/	0.37	Aweto (1981)
		3	/	0.43	
		7	/	0.71	
		10	/	0.71	
		OGF	/	0.71	

(continued)

Table 3.4 (continued)

Forest type	Soil	Age (yr)	Shannon diversity index	Simpson's index	Source
Tropical dry forest	/	<6	2.31	/	Ruiz et al. (2005)
		6–10	2.65	/	
		11–16	2.5	/	
		17–31	2.85	/	
		32–56	3.19	/	
		>56[c]	3.35	/	
	Lithosols and Haplic Phaeozems	1	1.36–3.34	/	Lebrija-Trejos (2008)
		3		/	
		5		/	
		7		/	
		10		/	
		12		/	
		14		/	
		18		/	
		22		/	
		27		/	
		32		/	
		37		/	
		40		/	
Tropical rain forest	Oxisols/Ultisols	1	1.96	0.59	Uhl (1987)
		2	2.33	0.68	
		3	2.09	0.65	
		4	2.57	0.71	
		5	2.86	0.73	

(continued)

Table 3.4 (continued)

Forest type	Soil	Age (yr)	Shannon diversity index	Simpson's index	Source
Lower montane forest/ Primary evergreen forest	Coarse sandy loamy soil	20–29	1.7–2.7 (median: 2.6)	/	Fukushima et al. (2008)
		20–29		/	
		20–29		/	
		20–29		/	
		20–29		/	
		30–39	1.8–2.2 (median: 1.9)	/	
		30–39		/	
		30–39		/	
		40–49		/	
		40–49		/	
		40–49		/	
		OGF	1.9–2.7 (median: 2.6)	/	
		OGF		/	
		OGF		/	
		OGF		/	
		OGF		/	
		OGF		/	
		OGF		/	

[a] Overstory
[b] Understory
[c] representing old-growth forest
[d] Sapling stratum (1–4.9 dbh)
[e] Tree stratum (dbh $\geq$ 5 cm)

as measures (Uhl 1987; Voeks 1996). The studied site was monitored from the first year after farm abandonment until year five of fallow. Both indices showed a general increasing trend from 0.59 to 0.73 for Simpson's index, and 1.96–2.86 for the Shannon index (from 1 to 5 years of fallow despite a slightly drop in the 3rd year; Table 3.4; Uhl 1987; Voeks 1996). This result is congruent with another chronosequence study carried out in the same region (Saladarriage et al. 1988), at least in the early fallow stage, where both demonstrated an increase in species diversity prior to 20 years of fallow (Table 3.3). Kammesheidt (1998) carried out another chronosequence study in the tropical moist forest of Paraguay and he observed a positive relationship between species diversity and fallow age. In the sapling stratum, the Shannon diversity index showed an increase from 1.79 in 2- to 5-year-old fallows to 2.33 in older fallows (10 and 15 years old). Levels recorded for these older fallows reached levels comparable with those of old-growth forest (2.34) (Kammesheidt 1998). While in the tree stratum, the Shannon diversity index increased from 2.14 to 2.84 when fallows aged, and the species diversity of the older stands almost approached the same value as the old-growth forest, which had a diversity of 3.09 (Kammesheidt 1998) (Table 3.4).

3.4.3 Species Diversity Decreases When Fallows Age

Although many studies reported an increasing trend in species diversity, a few studies have found a decrease in diversity as fallows aged. El-Sheikh (2005) concluded in his study that species diversity on fallow fields of dry forests in Egypt decreases when succession proceeds, despite an increase found in early stages (1–3 years). Another decreasing trend could be seen in the lower montane forest of northern Thailand, where the chronosequence of 20–49-year-old fallows was examined (Fukushima et al. 2008). The median value of the Shannon index decreased from 2.6 in 20–29-year-old fallows to 1.9 in the 30–49-year-old fallows. The Shannon index of 'uncultivated forests' was 1.9–2.7, with a median of 2.6, which is higher than the value of the oldest fallows reported in this study (Table 3.4; Fukushima et al. 2008). The authors suggested that the decrease might be due to the reduction in pioneer species found when fallows age, and the increase in sprouting species like *S. wallichii* on 30–49 year-old fields.

3.5 Discussion and Conclusions

Table 3.1 illustrates the diversity of results obtained when researchers examined the relationship between total species richness and fallow length. Some studies observed an increase when fallows age, while others found an increase at the beginning but then a drop in levels (or constant levels) in later stages. A few studies even concluded that there is no difference among age classes. For example,

Fig. 3.3 showed that total species richness in a study in Madagascar was not correlated with age class, and that a study in Bolivia found an increase in early age classes followed by constant levels in later age classes. No absolute pattern could be indentified even with the same type of forest. For instance, in Nigeria (Ross 1954) and Eastern Madagascar (Klanderud et al. 2010), studies examined tropical rain forest, but the changes in total species richness recovered at these study sites during succession were very different (Table 3.1). Ross (1954) found an increase when fallows aged, but Klanderud et al. (2010) did not. Therefore, we are not able extrapolate a general model for how total species richness is impacted by fallow duration.

Similarly, the time needed for total species richness to recover to levels found in old-growth forest samples also shows great variation, depending on the locality studied (Table 3.1). For example, the moist evergreen forest in Nigeria showed the fastest rate of total species richness accumulation, with the most comparable levels to that of the old-growth forest level. When looking at the other two tropical moist forests examined, in Bolivia and the Upper Rio Negro region of Colombia and Venezuela, the growth rate of total species richness also differed. Contrasting dry and moist forests, the growth rate of tropical dry forest in southern Mexico was found to be similar to that in the tropical moist forest in the Upper Rio Negro region, as both reached an equivalent level as that of the old-growth forest at around 40 years (Table 3.1). As there is great variation in results, it is not clear whether dry forests or humid forests posses a faster growth rate of total species richness. As seen from Table 3.2, an equivalent total number of species could be reached before 80 years, although some researches did not provide data on the old-growth forest (e.g. Breugel et al. 2007; Ross 1954; Raharimalala et al. 2010 and Fukushima et al. 2008). To determine the recovery rate of the fallows, it would be essential to include data from old-growth forest samples for comparison, as the climatic climax for each place is different. Thus, it would be difficult to conclude whether (and when) the secondary forest could recover to previous levels of undisturbed species richness.

Table 3.2 summarized the changes in the number of species from different studies. It would be difficult to conclude how total species richness and tree species richness are correlated with fallow length from the table alone, as variations are observed. However, as mentioned at the beginning of this section, the IDH predicts that species richness would peak at the intermediate successional stage (Ruiz et al. 2005). Ruiz et al.'s (2005) results did not agree with the hypothesis that (as mentioned above) total species richness shows a linear increase with fallow age. Yet, Raharimalala et al. (2010), Peña-Claros (2003) and Saldarriaga et al. (1988) found results that seem to agree with the IDH (Table 3.1). There seems to be great variations in the pattern of changes in species richness along the trajectory of succession.

More research is needed on the relationship between species richness and other environmental factors so as to better understand the forest regeneration that follows fallowing. Regression models like the Generalized Linear Model multiple regressions used in the study of tropical rainforest in Madagascar would allow researchers to examine how strong the relationship between fallow length (or other site factors) and species richness is, so as to present a clearer picture. Moreover, the rate at which species could reach similar richness as old-growth forest also shows variation in different places. The structural characteristics of the fallow field such as total basal area and total aboveground biomass are suggested to have an effect on the rate of species accumulation (Finegan and Delgado 2000), and studies on these aspects would reveal much.

Table 3.3 presents the summary of changes in total species diversity in different places. Again, a variation in changes in species diversity with regards to fallow age can be observed. Some studies found a constant increase, a few observed an increase in early stages of fallow but stable levels in later stages, while others reported a decreasing trend along succession. On the other hand, Table 3.4 summarized the values of the Shannon and Simpson's index from different studies. In general, the species diversity of early successional stages was low, as shown in the table, due to the dominance of a few pioneer grass species. But after those species died as succession proceeded, species diversity increased (Saldarriaga et al. 1988). Despite the fact that this phenomenon was found by most researchers, not all studies agreed (e.g. Fukushima et al. 2008). Thus, no single trend could be observed in the change in total species diversity during succession. Indeed, most results summarised in Table 3.3 do not agree with the intermediate disturbance hypothesis. Although Fukushima et al. (2008) found a decrease in species diversity in the intermediate fallow stage (after 20 years), there is a lack of information regarding the first 20 years of fallow (see Table 3.4). Thus, they were not able to predict the trend in the early fallow stages, and could not determine if their findings agreed with the intermediate disturbance hypothesis or not.

Studies that focused at the time needed to reach species diversity levels equivalent to those found in old-growth forest also found contrasting results. The years needed to restore old-growth forest levels ranged from seven to 80 (Table 3.3). The moist evergreen forest of Nigeria had the fastest species diversity growth rate (as well as the highest levels of species richness, as mentioned in the previous section). A comparatively slow growth rate was found in the tropical moist forest of the upper Rio Negro region, which required about 80 years to attain old-growth forest levels of species diversity. On the other hand, the tropical moist-forest of Bolivia required a much shorter time to reach a level of species diversity comparable to that of old-growth forests: only 10–30 years. Thus, different studies show considerable variation. For the two tropical dry forests in Colombia and southern Mexico, a similar period of time was needed to attain the levels of diversity of old-growth forest (both needed about 40–50 years). Whether dry forests or humid forests have faster species diversity growth rate is an open question, and we are unable to conclude one way or another due to the great variation in published results (see Table 3.4). Also, comparisons of the rates of recovery of fallows across studies might not be possible, as methods of measurement are different among studies—some researches adopted the Shannon diversity index while others used Simpson's index (see Table 3.4).

Both species richness and species diversity give important information as to the patterns of succession on the ageing fallows (Sanjit and Bhatt 2005). The research reviewed in this chapter shows that in some fields species richness and diversity peaks at intermediate stages while in other fields they increase as the fallow ages, while it others they decrease. The fact that results from different places show considerable diversity in that pattern of succession, implies that other factors, apart from the age of the fallows, are also very important in influencing succession. Some of these factors will be reviewed in the Chap. 5. A better understanding of the driving forces behind the processes of succession of species richness and diversity would be very helpful in improving the management of fallow fields. What is the impact of longer fallow periods on the patterns of change in species richness and diversity? What would be the 'appropriate' time for the fields to be left fallow before being cultivated again? Are longer fallow periods 'better' both for the land and the livelihood of the people? These are all unanswered questions which warrant further investigation.

References

Aweto AO (1981) Secondary succession and soil fertility restoration in south-western Nigeria. I, Succession. J Ecol 69:601–607

Bazzaz FA (1975) Plant species diversity in old-field succession ecosystems in southern Illinois. Ecology 56:485–488

Breugel MV, Bongers F, Martínez-Ramos M (2007) Species dynamics during early secondary forest succession: recruitment, mortality and species turnover. Biotropica 35:610–619

Brown S, Lugo AE (1990a) Tropical secondary forests. J Trop Ecol 6:1–32

Brown S, Lugo AE (1990b) Effects of forest clearing and succession on the carbon and nitrogen content of soils in Puerto Rico and US Virgin Islands. Plant Soil 124:53–64

Chazdon RL, Letcher SG, van Breugel M, Martínez-Ramos M, Bongers F, Finegan B (2007) Rates of change in tree communities of secondary Neotropical forests following major disturbances. Philos Trans Royal Soc B 362:273–289

Connell JH (1978) Diversity in tropical rain forests and coral reefs. Science 199:1302–1310

El-Sheikh MA (2005) Plant succession on abandoned fields after 25 years of shifting cultivation in Assuit, Egypt. J Arid Environ 61:461–481

Finegan B, Delgado D (2000) Structural and floristic heterogeneity in a 30-year-old Costa Rican rain forest restored, on pasture through natural secondary succession. Restor Ecol 8:380–393

Fukushima M, Kanzaki M, Thein HM (2007) Recovery Process of Fallow Vegetation in the Traditional Karen Swidden Cultivation System in the Bago Mountain Range, Myanmar. Southeast Asian Studies 45(3):303–316

Fukushima M, Kanzaki M, Hara M, Ohkubo T, Preechapanya P, Choocharoen C (2008) Secondary forest succession after the cessation of swidden cultivation in the montane forest area in Northern Thailand. For Ecol Manag 255:1994–2006

Hurlbert SH (1971) The non-concept of species diversity: a critique and alternative parameters. Ecology 52:577–586

Kammesheidt L (1998) The role of tree sprouts in the restoration of stand structure and species diversity in tropical moist forest after slash-and-burn agriculture in Eastern Paraguay. Plant Ecol 139:155–165

Kennard DK (2002) Secondary forest succession in a tropical dry forest: patterns of development across a 50-year chronosequence in lowland Bolivia. J Trop Ecol 18:53–66

Klanderud K, Mbolatiana HAH, Vololomboahangy MN, Radimbison MA, Roger E, Totland Ø, Rajeriarison C (2010) Recovery of plant species richness and composition after slash-and-burn agriculture in a tropical rainforest in Madagascar. Biodivers Conserv 19:187–204

Lebrija-Trejos E, Bongers F, Pérez-García EA, Meave J (2008) Successional change and resilience of a very dry tropical deciduous forest following shifting agriculture. Biotropica 40:422–431

Lloyd M, Ghelardi RJ (1964) A table for calculating the "equitability" component of species diversity. J Anim Ecol 33:217–225

Margalef R (1958) Information theory in Ecology. Gen Syst 3:36–71

Margalef R (1968) Perspectives in ecology theory. University Chicago Press, Chicago

McIntosh RP (1967) An index of diversity and the relation of certain concepts to diversity. Ecology 48:392–402

Metzger JP (2003) Effects of slash-and-burn fallow periods on landscape structure. Environ Conserv 30(4):325–333

Naveh Z, Whittaker RH (1979) Structural and floristic diversity of shrublands and woodlands in north Israel and other Mediterranean areas. Vegetatio 41:171–190

Peña-Claros M (2003) Changes in forest structure and species composition during secondary forest succession in the Bolivian Amazon. Biotropica 35(4):450–461

Pielou EC (1966) Species-diversity and pattern-diversity in the study of ecological succession. J Theor Biol 10(2):370–383

Raharimalala O, Buttler A, Ramohavelo CD, Razanaka S, Sorg JP, Gobat JM (2010) Soil-vegetation patterns in secondary slash and burn successions in Central Menabe. Madag Agric Ecosyst Environ 139:150–158

Ross R (1954) Ecological studies on the rain forest of southern Nigeria. III, Secondary succession in the Shasha reserve. J Ecol 42:259–282

Ruiz J, Fandiño MC, Chazdon RL (2005) Vegetation structure, composition, and species richness across a 56-year chronosequence of dry tropical forest on Providencia Island. Colombia, Biotropica 37(4):520–530

Saldarriaga JG, West DC, Tharp ML, Uhl C (1988) Long-term chronosequence of forest succession in the upper Rio Negro of Colombia and Venezuela. J Ecol 76:938–958

Sanjit L, Bhatt D (2005) How relevant are the concepts of species diversity and species richness? J Biosci 30(5):557–560

Shannon CE, Weaver W (1949) The mathematical theory of communication. University of Illinois Press, Urbana

Sheil D (2001) Long-term observations of rain forest succession, tree diversity and responses to disturbance. Plant Ecol 155:183–199

Sheil D, Burslem D (2003) Disturbing hypotheses in tropical forests. Trends Ecol Evol 18:18–26

Simpson, EH (1949) Measurement of diversity. Nature 163:688

Simpson GG (1964) Species density of North American recent mammals. Syst Zool 13:57–73

Spellerberg IF (1991) Monitoring ecological change. Cambridge University Press, Cambridge

Toledo M, Salick J (2006) Secondary succession and indigenous management in semideciduous forest fallows of the Amazon Basin. Biotropica 38(2):161–170

Tran P, Marincioni F, Shaw R (2010a) Catastrophic flood and forest cover change in the Huong river basin, central Viet Nam: a gap between common perceptions and facts. J Environ Manag 91(11):2186–2200

Tran VD, Osawa A, Nguyen TT (2010b) Recovery process of a mountain forest after shifting cultivation in Northwestern Vietnam. Ecol Manag 259:1650–1659

Tran VD, Osawa A, Nguyen TT, Nguyen BV, Bui TH, Cam QK, Le TT, Diep XT (2011) Population changes of early successional forest species after shifting cultivation in Northwestern Vietnam. New Forest 41:247–262

Uhl C (1987) Factors controlling succession following slash-and-burn agriculture in Amazonia. J Ecol 75:377–407

Voeks RA (1996) Tropical forest healers and habitat preference. Econ Botany 50(4):381–400

Chapter 4
Species Composition

Abstract Chapter 4 reviews research on species composition: how vegetation shifts from grasses to trees in different successional stages. It is generally accepted that a transition from grasses to herbs to shrubs to woody species is common, despite the fact that time of recovery may vary in different places. Also, it is usually suggested that with a shorter fallow, tree growth would be hindered, and the land might not be able to reach a tree fallow. We also discuss what different researches have found regarding the relationship between soil depletion and the reduction of tree species as fallow duration shortens. Whether the changes in species composition are predictable is one of the major questions regarding secondary forest succession. Two influential models have been proposed to predict floristic change: the 'relay floristics' model and the 'initial floristic composition' model. However, our review concludes that new approaches are needed to explain the great variability of species composition among sites, as neither the 'reductionist' nor the 'holistic' approaches seem to be able to fully predict and explain the regeneration pathway of species composition along the successional trajectory.

Keywords Ecological succession · Forest structures · Species composition · 'Relay floristics' model · 'Initial floristic composition' model · Pioneer species

4.1 Introduction

During the successional process following shifting cultivation, there are not only changes in the forest structures, but also in the composition of species. These changes of species composition involve the transition of species to different successional stages and the different patterns of recruitment after land abandonment. When looking at the types of vegetation that evolve during succession, it is

C. O. Delang and W. M. Li, *Ecological Succession on Fallowed Shifting Cultivation Fields*, SpringerBriefs in Ecology, DOI: 10.1007/978-94-007-5821-6_4, © The Author(s) 2013

generally accepted that a transition from grasses to herbs to shrubs to woody species is common, despite the fact that time of recovery may vary in different places. For example, in humid tropical areas, a few studies have described how fallow transitions directly from grassland to rainforest (Humbert 1927; Lowry et al. 1997; FAO 2000; Styger et al. 2007). Also, it is usually suggested that with a shorter fallow, tree growth is hindered, and the land might not be able to reach a tree fallow (Neba 2009; Styger et al. 2007). In this chapter, we look at how vegetation shifts from grasses to trees in different successional stages, and discuss the findings of various studies that report different times to reach a tree fallow.

Whether the changes in species composition are predictable is one of the major questions regarding secondary forest succession (Chazdon et al. 2007). A large number of studies have been conducted to examine the changes in species composition along succession trajectories, with a diversity of results. Two influential models to predict floristic change have been proposed by Egler (1954). These models are the 'relay floristics' model and the 'initial floristic composition' model (Chazdon et al. 2007). The idea of the 'relay floristics' model is that forest succession proceeds through a wave of colonization by different species; in other words, a sequence of changes in species composition occurs, which is predictable (Chazdon et al. 2007). On the other hand, the 'initial floristic composition' model hypothesises that most species colonise together, including pioneer and shade-tolerant species, at the very beginning of fallow, but dominate at different successional stages (Fig. 4.1) (Chazdon et al. 2007; Breugel et al. 2007). After canopy closure, pioneers die out while shade-tolerant species continue to occupy the region (Breugel et al. 2007).

In Fig. 4.1, species one (1) through five (5) establish at the initiation of succession, and successively dominate the site due to the different lengths of time they take to reach maturity. Species one may represent an annual herb, species four a short-lived shrub, and species five a slow-growing, shade-tolerant forest tree. Species one to four fail to regenerate as they are light-demanding. Late colonization

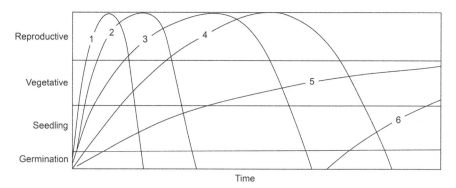

Fig. 4.1 Schematic representation of the initial floristic composition model of succession. *Source* Finegan (1984)

by species six occurs as a consequence of the release of the inhibitory effects upon it of species three (modified from Gómez-Pompa and Vásquez-Yanes 1981).

To explain the dynamic of succession and species recruitment, two other theories have been proposed: the holistic theory and the reductionist theory (Mitja et al. 2008; Klanderud et al. 2010; Finegan 1984). The holistic model suggests that pioneer species establish first in early successional stages, and then forest species invade at later stages (Mitja et al. 2008). The idea is that pioneer species are able to facilitate the establishment of forest species (Mitja et al. 2008); thus, this hypothesis is often called the 'facilitation' hypothesis, as it assumes that a suitable niche will be created by a particular group of pioneer species for the next group to utilize (Finegan 1984). Under this holistic approach, ecological succession is characterized by 'wave-like invasions by groups of species' (Finegan 1984 p. 109).

On the other hand, the reductionist theory is based on the 'initial floristic composition' model (Finegan 1984). This theory predicts that pioneer trees will be replaced by forest species as the latter grow. This is because light will be blocked by growing plants, and the seedlings of pioneer species will not be able to survive under the shade generated by the forest species (Finegan 1984; Knight 1975; Saldarriaga et al. 1988). In this chapter, we review the results presented in different studies, in terms of these two models.

4.2 Succession from Grasses to Trees

A study conducted in Madagascar, describes how ecological succession is locally divided into 3 main stages, *Ramarasana*, *Dedeka* and *Savoka* (Styger et al. 2007). *Ramarasana* refers to the early grass fallow stage where crops are just harvested and then the land is quickly overgrown by herbaceous plants. According to local farmers, this stage can take 6 months to 2 years before the second stage, *Dedeka*, is reached. *Dedeka* is a stage characterized by shrubby fallow; this stage is dominated by short (height ∼1–1.5 m) shrubby species. As the succession proceeds, more and more woody species replace the herbaceous plants until the fallows reach the stage of *Savoka*, where shrubs are taller (2–4 m) and with bigger stem diameters. The fallow will reach a stage of tree fallow (*Savoka Mody's*) when the land is left to further regenerate. Research conducted in the Brazilian Amazon concluded that for fallow fields that are more than 5 years old, the mean patch size is 2–5 fold higher, and the mean cover area of the forest 2–3 fold higher than short fallows (under 5 years old) (Metzger 2003). In lowland moist and wet Neotropical regions, Guariguata and Ostertag (2001) concluded that in the first 10 years of fallow, vegetation is generally dominated by grasses and shrubs. Then, pioneer trees species invade the fields. By later stages, taller and longer-lived trees account for the biggest community. On the other hand, Lebrija-Trejos et al. (2008) also described vegetation changes throughout the succession process following shifting cultivation in a tropical dry forest of southern Mexico. According to their data, the proportion of shrubs decreases with fallow age from more than 80 % during the first 3 years to 59 % in older fallows. Shrubs species like *Waltheria indica*, *Chamaecrista nictitans* var. *jaliscensis* and *Melochia tomentosa* dominated during 1–3 years-old fallows, and accounted for 68–82 % of all of the woody species found on the field (Lebrija-Trejos et al. 2008). After 3 years, Lebrija-Trejos reported a domination of pioneer trees such as *Mimosa tenuiflora* and *M. acantholoba* var. *eurycarpa*, wherin more than 95 % of the total basal area was shared by *Mimosa tenuiflora* in 3–5 years old fallows. In contrast, *Mimosa acantholoba* dominaed after 7 years. The forest species *A. adstringens, A. paniculata, E. schlechtendalii, L. divaricatum* and *Senna atomaria* started to emerge after 32 years as they were able to grow under the canopy of the pioneer *Mimosa* species (Lebrija-Trejos et al. 2008).

The domination of grass and shrubs in the early successional stage could be explained by the following scenario: it is likely that light seeds that spread by wind regenerate easily as they are abundant, able to reach greater distances and readily penetrate through soil (El-Sheikh 2005). In contrast, the seeds of woody species tend to be larger and fewer in number, thus less likely to disperse and colonize areas at greater distances (El-Sheikh 2005). As a result, the early fallow stage is dominated by a large number of weedy species (El-Sheikh 2005). However, the length of domination by grass and shrubs on a fallow site varies; as shown above, shrubs dominated for a shorter period of time in the dry forest of southern Mexico (Lebrija-Trejos et al. 2008) in comparison to Neotropical regions (Guariguata and Ostertag 2001).

4.2.1 Location-Specific Patterns of Change

Despite the fact that the change of vegetation type along the successional trajectory is generally the same in different parts of the world, the rate of change shows great variation, even within a geographical region. For example, a number of studies have focused on succession in the forests of Madagascar, finding that the time to reach a fallow stage dominated by trees varies (Styger, et al. 2007; Klanderud et al. 2010; Raharimalala et al. 2010; Von Schulthess 1990; Lebrija-Trejos et al. 2008). A chronosequence study along the south-western coast of Madagascar (Raharimalala et al. 2010) examined the time needed for trees to regenerate following fallow. The results of this study agreed with one in which researchers collected data by interviewing farmers in the rainforest villages of eastern Madagascar (Styger et al. 2007). Both studies came up with similar results: woody species appeared after 5 years of fallow, while tree species dominated in 20-years-old fallows. Schmidt-Vogt (1998) found a similar rate of woody species establishment in northern Thailand where after 3 years woody plants increased in number and trees developed after 8 years of fallow, when most weeds had already died out. Another chronosequence study of tropical rainforests in Eastern Madagascar (Klanderud et al. 2010) examined 33 fallows ranging in age from 1 to 26 years post-shifting cultivation, and found that trees reach a similar level of coverage than shrubs after 26 years of fallow.

These results were similar to those observed by Raharimalala et al. (2010). However, Von Schulthess (1990) found that in the drier part of western Madagascar, after 25–30 years of fallowing the field is only covered with mature shrub. About 50–60 years are needed for the vegetation to develop into a secondary forest at this site. The differences observed between studies in Madagascar might be due to the variation in climate, topography, soil conditions or other site-specific characteristics among regions. This would agree with Raharimalala et al.'s (2010) claim that according to the FAO (2003), the main factors that cause differences in the succession process include climate, altitude and soil type.

In the dry forest of Egypt, after examining 39 fallow sites ranging in age from 1 to 3 years, 5 to 6 years, and 25 years, El-Sheikh (2005) concluded that perennials and species with grassy-woody characteristics begin to dominate in year 5–6, while woody and grassy species become the major type of vegetation after 25 years of fallow. These findings are similar to those of Raharimalala et al. (2010), Styger et al. (2007) and Klanderud et al. (2010). A similar pattern could also be found in Sudan (Halwagy 1963). On the other hand, a chronosequence study conducted in the mixed deciduous forest of central Burma (Myanmar) found a faster rate of tree establishment. In this study, trees start increasing in number in the fifth year of fallow and dominate in the 15th year (Fukushima et al. 2007). Uhl (1987) did a long-term, permanent-plot study in the upper Rio Negro region of Amazon where they monitored fallow vegetations from the first through the fifth year of abandonment. The author observed a quicker colonization of woody species from the second and third year of fallow following the recruitment of forbs and grasses during the first year. In addition, Rouw (1993) conducted a permanent-plot study of fields with 3 years of cultivation and 2 years of fallow in the rainforest of the south-western region in Côte d'Ivoire. In this region, pioneer trees were well established after 6 months of land abandonment following one rice harvest, and formed a close canopy with a height range of 2–4 m (Rouw 1993). This example showed an even faster recruitment of woody species compared to fallows in the upper Rio Negro (Uhl 1987).

In short, the rate of woody species establishment during the process of succession can be very different depending on location (Table 4.1). It is suggested that other factors like land use history and intensity play an important role in the successional process (Williams-Linera et al. 2011; Miller and Kauffman 1998; Molina Colón and Lugo 2006; Romero-Duque et al. 2007), which will be discussed in a later section.

4.3 Fallow Length and Species Composition

Some researches have used correspondence analysis to evaluate how environmental variables relate to species composition. A study carried out in the tropical moist forest of Bolivia found that species composition varied with stand age, with the greatest variation reported in the canopy layer (Peña-Claros 2003). The correspondence analysis was used to evaluate how fallow age affects species

Table 4.1 Time needed for tree establishment in fallow, by location

Place	Forest type	Rate of trees regeneration	Source
South-western coast of Madagascar	Dry deciduous forest	Appeared after 5 years; dominated after 21–30 years	Raharimalala et al. (2010)
Eastern Madagascar	Rainforest	Appeared after 5–6 years; dominated after 15–20 years	Styger et al. (2007)
Eastern Madagascar	Tropical rainforest	Dominated after 26 years	Klanderud et al. (2010)
Western Madagascar	/	Just mature shrub after 25–30 years	Von Schulthess (1990)
Egypt	Dry forest	Dominated after 25 years	El-Sheikh (2005)
Central Burma	Mixed deciduous forest	Dominated after 15 years	Fukushima et al. (2007)
Upper Rio Negro region	Tropical rainforest	Appeared after 2–3 years	Uhl (1987)
South-western Côte d'Ivoire	Rainforest	Appeared after 6 months	Rouw (1993)
Northern Thailand	Evergreen montane forests	Appeared after 8 years	Schmidt-Vogt (1998)
Southern Mexico	Tropical dry forest	Dominated after 3 years	Lebrija-Trejos et al. (2008)

composition, and after examining 250 species, Peña-Claros concluded that the major variable that causes the difference in species composition is the age of the fallow. Despite this, locational differences of the two sample sites in the study (El Turi and El Tigre) did contribute to the variation to a certain extent (Peña-Claros 2003). Another Canonical Correspondence Analysis (CCA) was conducted in Eastern Madagascar, and environmental variables such as slash and burn cycle and years since last abandonment of the fallow were taken into account to evaluate significance with respect to species composition (Klanderud et al. 2010). Results showed that tree seedling, tree sapling and shrub species composition were significantly correlated with fallow duration, while the species composition of adult trees was found to have no significant relationship with fallow length; that is, no significant change in numbers of adult trees was recovered between the youngest fallows and old-growth forests (Klanderud et al. 2010).

Researchers report that along the successional process, different stages are dominated by different species. For example, a total of 141 usable plant taxa were found in 1–11 years-old fallows in the forest of northern Lao P.D.R. (Delang 2007). Most of these plants only appeared in a certain age class (e.g. two species were only found in young fallows of 1–3 years old, while six species were only found in fallows of 5 years of age or older). Only 14 plants were indentified throughout the 11 years fallow period (Delang 2007). A chronosequence study conducted in north-western Vietnam (Tran et al. 2010a, b) and the Bolivian lowland (Toledo and Salick 2006) tested the relative importance of a variety of species on each fallow phase. In north-western Vietnam, relative stem density (RD) among species showed that no single species dominates the whole process of succession (Tran et al. 2010a, b). For example, *Eurya trichocarpa* (Theaceae) had an RD of 14 % during 7–10 years of fallow, demonstrating that it dominated this fallow stage. In contrast, *Choerospondias axillaris* (Anacardiaceae) dominated during years 18 to 26, with RD values of 24 % and 13 %, respectively (Tran et al. 2010a, b). Toledo and Salick (2006), using the Importance Value Index (IVI) as a measure, also found that different fallow stages are dominated by different species. Therefore, the variation in species composition is clearly related to stand age. How species composition changes along succession will be discussed in the following sections in more detail.

4.3.1 Similarity of Species Composition in Similar-Aged Fallows

Although species' responses to succession are deemed to be highly individualistic (Toledo and Salick 2006), a similarity among age classes was found by a number of researches. The Bray-Curtis ordination was used in a study by Toledo and Salick 2006 to determine the species composition in a chronosequence study of a Bolivian forest. In the overstory, authors reported a separation of the early stands from the other fallows due to a large amount of *Trema micrantha* in young fallows. In contrast, other fields with intermediate age classes were found to be closer in species composition (Toledo and Salick 2006). Similarly, the species composition of the oldest fallows (36 years old) was different from the fallows in other stages, and had species similar to those in old-growth mature forest (Toledo and Salick 2006). These species included *Pseudolmedia laevis* and *Licaria triandra* (Toledo and Salick 2006). In the understory, a similar pattern was observed where species indentified in the old-growth forest were very different from species found in young fallows (Toledo and Salick 2006). Species such as *Iresine diffusa*, *Axonopus compressu* and *Conyza bonarienses* were the most abundant in young fallows, while *Bolbitis serratifolia* and *Piper aleyreanum* were the dominant species in old-growth forests (Toledo and Salick 2006). Likewise, in a study conducted in Mexico, Breugel et al. (2007) examined the similarity in species composition of 1–5 years-old fallows before and after 18 months using the Chao-Jaccard abundance-based similarity estimator. The value of this estimator ranged from 0.95 to 1 in all 1–5 years-old fallow plots after 18 months (1 = completely similar) (Table 4.2). This data indicates that species composition remained more or less the same at this site after 18 months, which could be due to the high dominance of a few species. This may have reduced the numerical importance of newly recruited species (Breugel et al. 2007). Moreover, although a total of 60 % of the species found on fallows between 1 and 5 years of age could be indentified in the old-growth forest, the five species that dominated on the fallows (66 % of all trees found) accounted for less than 1.3 % of the total number of trees in the old-growth

Table 4.2 Similarity between species assemblage of censuses one (C1) and two (C2) and between recruitment (R), dead trees (M), and surviving trees (S) assemblages. Chao-Jaccard abundance-based similarity estimators are given ± standard error. Bold values indicate that the value is significantly different from one. The number in the plot names indicates plot age at census one

Plot	C1*C2	S*M	S*R	M*R
R1	0.98 ± 0.04	0.90 ± 0.14	0.83 ± 0.16	0.35 ± 0.16
F2	0.95 ± 0.05	0.87 ± 0.10	0.86 ± 0.15	0.74 ± 0.23
H2	1.00 ± 0.03	0.83 ± 0.17	0.96 ± 0.10	0.78 ± 0.27
P2	0.98 ± 0.04	0.83 ± 0.14	0.94 ± 0.14	0.73 ± 0.30
R2	0.97 ± 0.03	0.99 ± 0.07	0.54 ± 0.14	0.57 ± 0.20
G3	1.00 ± 0.00	0.98 ± 0.13	1.00 ± 0.06	0.85 ± 0.28
F4	1.00 ± 0.01	0.93 ± 0.17	0.78 ± 0.19	0.38 ± 0.21
R5	1.00 ± 0.01	0.97 ± 0.08	0.39 ± 0.13	0.39 ± 0.20

Source Breugel et al. (2007), Table 1

forest (Breugel et al. 2007). This indicates that the species composition in young fallows is clearly different from that of old-growth forests.

Aweto (1981) investigated how similar the species composition of fallows was with respect to age, and included the old-growth forest by ordination. Aweto found that a distinct pattern of species composition existed between fallows and old-growth forest in Nigeria: the 40 fallow plots cluster in the centre, while the 10 old-growth forests group below (plots 41–50; Fig. 4.2). Figure 4.3 describes the other ordination for the 40 fallow plots, and it demonstrates that one-year-old plots (numbers 1–10) cluster near the centre, but 10-years-old plots (numbers 31–40) group on the right side of the ordination (Fig. 4.3). The result indicates that species

Fig. 4.2 Principal components analysis of species composition in 50 (fallows and forest) plots. *Source* Aweto (1981)

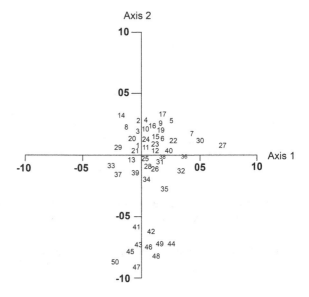

Fig. 4.3 Principal components analysis of species composition in forty fallow plots. *Source* Aweto (1981)

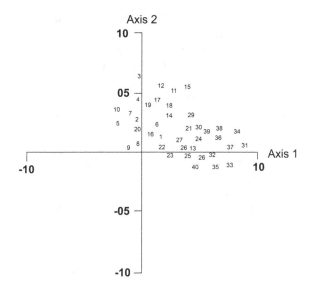

composition of old-growth forest, 10-years-old fallows and one-year-old fallows are all different (Aweto 1981). For instance, trees species such as *Bosqueia angolensis, Celtis brownie, Cola gigantean* and *Pentaclethra macrophvlla* were only recorded in the mature forest. On the other hand, in 10 years-old plots, woody species like *Allophvllus africanus, Anthonotha macrophylla, Ficus exasperate* and *Harungana madagascariensis* dominate (Aweto 1981). However, Aweto (1981) pointed out that the woody species that occur in the 10-years-old fallows were also found in 7 years-old plots, while species composition on 1 year-old fallows was similar to that found on 3 years-old fallows. This shows that the species composition of fallows of similar ages are more identical than those of more different ages.

4.3.2 Pioneer Species Recruit First During Succession

A number of studies found that pioneers were the first to colonize the fallow site after abandonment. The chronosequence study carried out in the upper Rio Negro region found that pioneers colonized the fallow fields in early successional stages from the first 10 to 20 years, but that from 30 to 40 years onwards, the pioneers died out and were replaced by longer-lived species like *Vochysia sp., Alcornea sp.* and *Jacarajda copaia*. These species' dominance could last for up to 50 years in the presence of ideal conditions (Sladarriaga et al., Saldarriaga et al. 1988). Tran et al. (2011) reported that only four pioneer species established during the first year of fallow (*Wendlandia paniculata, Schima wallichii, Camellia tsaii* and *Lithocarpus ducampii*), and dominated at 10 years. At this time, these four species accounted for 78 % of the total species surveyed. However, these species decreased to 9 % of the total species surveyed from years 21 to 30, and eventually disappeared completely in the old-growth forest (Tran et al. 2010a, b, 2011). This result is congruent with a hypothesis proposed by Osunkoya et al. (1994) and Walters and Reich (1996), which states that shade-intolerant species dominate in the early successional stage but are limited to regions with a large amount of light, and thus later decrease in abundance as shade increases.

To see how the plants' growth is affected, Osunkoya et al. (1994) created a lighting environment similar to that of rainforests in a study site in northern Queensland, Australia. Their results showed that species with larger seeds grow into larger seedlings, which show less sensitivity to shade than species with small seed sizes. Similarly, Walters and Reich (1996) conducted an experiment with controlled light conditions. Under 8 % of total light, small-seeded shade intolerant species like *B. papyrifera,* and mid-tolerant species such as *B. alleghaniensis,* had higher growth rates than the shade-tolerant ones (*Ostrya* spp. and *Acer* spp.). However, under 2 % total light, the growth for the shade intolerant species was lower than that of the shade-tolerant *Ostrya* spp. and *Acer* spp.: about 50 % of *B. papyrifera* and *B. alleghaniensis* individuals died in 2 % light conditions (Fig. 4.4;

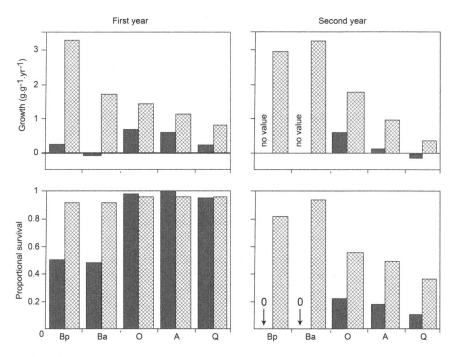

Fig. 4.4 Growth (proportional change in biomass) and proportional survival for the first and second growing seasons in 2 % (*black bars*) and 8 % (*gray bars*) light. Species are ordered (*left* to *right*) according to increasing initial seedling mass and seed mass (Walters et al. 1993) and decreasing relative growth rate in high light (Walters et al. 1993. Species acronyms are *Bp* = *Betula papyrifera*, *Ba* = *Betula alleghaniensis*, *O* = *Ostrya virginiana*, *A* = *Acer saccharum*, and *Q* = *Quercus rubra*. *Source* Walters and Reich (1996)

Walters and Reich 1996). These experiments demonstrated that the shade-intolerant species dominate in the beginning of succession as they can only grow in an environment with a sufficient amount of light (8 %; Tran et al. 2011).

4.3.3 Both Pioneer and Forest Species Recruit Together in the Early Stage of Succession

While the above examples suggest that pioneer species are the first to establish on fallows, different successional patterns were found by different researchers. A number of studies report that pioneer and shade-tolerant forest species are found at the beginning of succession, although they dominate during different stages of fallow (Uhl 1987; Peña-Claros 2003; Chazdon et al. 2007; Breugel et al. 2007; Guariguata and Ostertag 2001; Williams-Linera et al. 2011; Klanderud et al. 2010).

In the chronosequence study conducted in the tropical rainforest of eastern Madagascar, Klanderud et al. (2010) found that both pioneer and forest species established together at the very beginning of succession. Canonical Correspondence Analysis found that seedlings of forest tree species like *Albizia chinensis*, *Croton mongue*, *Ficus baronii*, *Harungana madagascarensis* and *Tambourissa lastelliana* were present in the youngest fallows, where they established quickly once land was abandoned. Peña-Claros (2003) found that pioneer and forest species colonized in the early stage but show dominance in different stages depending on their life spans and rates of growth. The author reported the presence of late successional species (such as *Euterpe precatoria*, *Eschweilera coriacea*, and *Pseudolmedia laevis*) whose abundance levels were positively associated with fallow age in the early fallows (2–3 years old). However, some species established only 20 years after land abandonment (Peña-Claros 2003).

Three other researches found an earlier appearance of forest species on fallows compared to Peña-Claros' results in Bolivia. These studies were conducted in the lowland tropical rain forest of Mexico (Breugel et al. 2007), the rain forest of the upper Rio Negro region (Uhl 1987) and the tropical dry forest of the eastern lowlands of Bolivia (Kennard 2002). These studies reported the recruitment of shade-tolerant forest species during the first year of fallow. For example, Breugel et al. (2007) examined eight abandoned cornfields 1–5 years after they were last cultivated, and recorded both pioneer and shade-tolerant species at the beginning of succession. Shade-tolerant tree species accounted for 2 % of all species on 1-year-old fallows, but after 18 months, shade-tolerant species made up 34 % of the total number of species. Despite this large increase in proportion of shade-tolerant species, the authors remarked that the incidence of shade-tolerant species was highly variable among fallows, perhaps due to the differences in the loss of basal area in the plots (Breugel et al. 2007). These same authors found that the proportion of pioneer species' recruitment is positively associated with the percentage of basal area mortality, which decreased from 22 to 74 % due to the death of canopy trees.

In the upper Rio Negro study conducted by Uhl (1987), a total of five forest species were recorded in the first year of fallow. These species included the taxon *Goupia glabra* Aublet, which accounted for 2.5 % of the total number of plants found on the plot (Uhl 1987). At the same time, the author reported a domination of pioneer tree species on the field. These pioneer species accounted for 71 % of the total tree species found. A chronosequence study carried out in lowland Bolivia by Kennard (2002) examined 14 fallows with 12 different age classes ranging from 1 to 50 years. Kennard tested the domination of short-lived pioneers, long-lived pioneers, partially-shade-tolerant and shade-tolerant species by their fraction of total basal area. The study showed that both long-lived pioneers, partially-shade-tolerant and shade-tolerant species established together during the first year of fallow, with long-lived pioneers as the most abundant guild (Fig. 4.5; Kennard 2002).

Another chronosequence study carried out in the tropical dry forest of Mexico selected five fallow sites ranging from 7 months to 6 years in age, all with different land use history (Williams-Linera et al. 2011). The results showed that in the youngest fallow of 7 months, primary species (shade tolerant species that only grow in the tropical dry forest) already accounted for a considerable proportion of

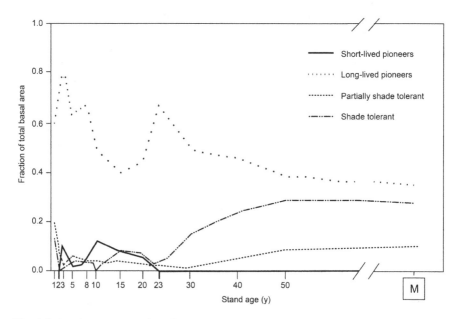

Fig. 4.5 Dominance (proportion of total basal area) of different regeneration guilds (short-lived pioneers, long-lived pioneers, partially shade-tolerant and shade-tolerant). The graph shows a 50-y chronosequence of forest fallows following agricultural abandonment. Classification of species into regeneration guilds follows Pinard et al. (1999). The mature forest is represented by the letter 'M'. *Source* Kennard (2002)

the total species surveyed. By 8 months, primary species became the largest proportion when compared to secondary (pioneer tree species that are shade intolerant and grow in deforested areas) and intermediate (species that naturally appear in dry forest but also establish on fallows with intermediate stage) species (Fig. 4.6). A total of 10 % of the intermediate and primary species that could be indentified in the old-growth forest was already recruited in this early succession stage. These species included *Brosimum alicastrum*, *Malpighia glabra*, *Randia aculeata*, *Spondias purpurea*, *Tabebuia chrysantha* and *Trichilia trifolia* (Williams-Linera et al. 2011). Moreover, the authors reported that 20 species could be found in both early successional sites and old-growth mature forest, but that the importance value index showed that none of these species had a high frequency in both fallows and old-growth forest (Fig. 4.7).

In short, some studies concluded that pioneers recruited first in the beginning of succession, while other studies reported that both pioneers and shade-tolerant forest species established simultaneously. It is possible that both patterns may be observed in the tropical areas, and that patterns depends more on the intensity of human disturbance than on qualities specific to forest type (Uhl 1987). The affect

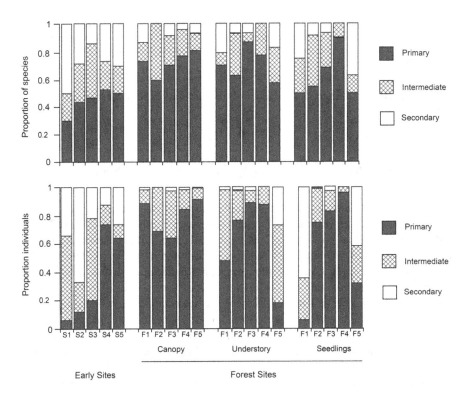

Fig. 4.6 Proportion of species and trees (individuals) classified by successional status. At early successional sites (S) and forests (canopy, understory and seedlings) in the tropical dry forest region of central Veracruz, Mexico *Source* Williams-Linera et al. (2011)

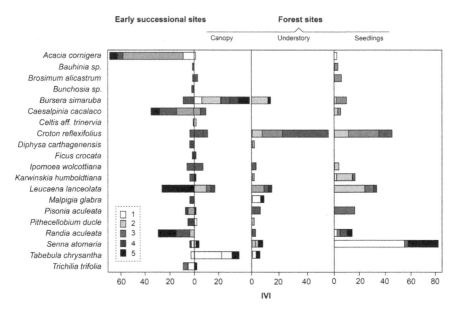

Fig. 4.7 The importance value index of tree species recorded at both early secondary successional sites and forests (canopy, understory and seedlings) The importance value index is given for each species, and numbers correspond to the successional and nearby forest sites *Source* Williams-Linera et al. (2011)

of the intensity of human disturbance on recruitment will be discussed in later sections.

4.3.4 Replacement of Pioneers by Shade-Tolerant Species

Studies in the literature also report the replacement of pioneer species by shade-tolerant species during succession, but with different patterns observed in different studies. As mentioned previously, Tran et al. (2010a, b, 2011) found that pioneer species disappeared when fallows aged in north-western Vietnam. The authors suggested that the replacement of pioneers might be due to the lack of available sunlight and the elimination by 'obligation succession' (Horn 1976; Tran et al. 2010a, b, 2011). Under this model, pioneer species establish first on fallows, and other new species are recruited under their shade. Subsequently, pioneers disappear as shade is created, because the canopy starts to close and restricts sunlight that they need to grow (Tran et al. 2010a, b, 2011). It is suggested that at a later stage, more shade-tolerant species grow as those pioneer trees have created an ideal microenvironment within which they can germinate (Tran et al. 2011). For example, the shade-tolerant species *Ficus racemosa* and *Celtis sinensi* were only found after 26 years of fallow (Tran et al. 2010a, b).

A later domination of forest species was observed in the upper Rio Negro region where after between 60 and 80 years, forest species (e.g. *Mouriri uncitheca* Morley and Wurdack, *Clathrotropis* sp and *Couma utilis* (Mart) Muell-Arg) started to appear as saplings while the longer-lived species that previously dominated the field died (Saldarriaga et al. 1988). However, although some pioneers died out when fallows aged (and were replaced), some other pioneer species like *Cecropia* sp, *Miconia myriantha* Bentham and *Vismia japurensis* persisted in larger gaps in the field (Saldarriaga et al. 1988). Thus, unlike the findings reported in northwestern Vietnam, which found that pioneers would be wiped out eventually by shade-tolerant species, Saldarriaga et al. (1988) demonstrate that it is possible for pioneer species to persist. Similarly, Breugel et al. (2007) reported that the replacement of pioneers by shade-tolerant species occurs in the early stage of succession, not later. Referring to Figs. 4.8a, b, the shade-tolerant fraction (the fraction of total number of trees in the sample that belongs to shade-tolerant species) shows that species recruited after 18 months had a higher shade-tolerant fraction than the species that survived and died, although great variation was shown among plots in the study. Nevertheless, the substantial recruitment of pioneer species could be found on some of the plots. For instance, the pioneer species *Trichospermum mexicanum* had a high level of recruitment, which lead to an absolute increase of 260 and 126 % in two different plots. This high increase was due in part to the species' recruitment and the disappearance of the dominant pioneer species in the area: *Trema micrantha* (mortality rate of 89 and 93 %; Breugel et al. 2007). However, these authors did not find this phenomenon in all of

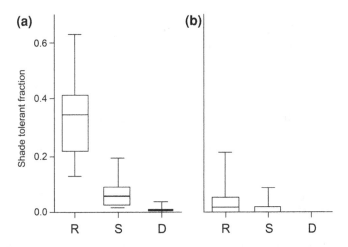

Fig. 4.8 Dynamics of the fraction of shade-tolerant trees in early successional communities. Shade tolerant fraction of the trees of the initial assemblage that survived the study period (S) and of the groups of trees that recruited (R) and that died (D) between censuses one and two. Different lower size limits were used to define the initial group of trees as well as recruitment: **a** height ≥ 1.5 m; **b** dbh ≥ 2.5 cm. Boxes show median, the 25th and 75th percentiles and the whiskers give the extent from lowest to highest values. *Source* Breugel et al. (2007)

the plots surveyed, so changes in species composition are highly dynamic and vary spatially.

On the other hand, a study conducted in the tropical dry forest of Bolivia failed to find the replacement of pioneer species by shade-tolerant species (Kennard 2002). This study reported that long-lived pioneers dominate from the youngest to the oldest fallows (1–50 years old) and are not replaced. Although shade-tolerant species start dominating in older plots and their importance value (relative abundance + relative dominance + relative frequency)/3) increases, they are not able to dominate total basal area, even in the mature forest (Fig. 4.5). As a result, the authors did not observe the replacement of pioneers by shade-tolerant species.

Comparison between studies is difficult, as different methods were used to evaluate the species' replacement with succession, and the differentiation of pioneers and shade-tolerant species were also not the same. For instance, Breugel et al. (2007) performed a Detrended Correspondence Analysis (DCA) to evaluate the species turnover when succession proceed, and pioneers and shade-tolerant species were put into groups of Recruited, Dead or Surviving trees. On the other hand, Kennard (2002) evaluated the dominance of species groups by looking at the proportion of total basal area they comprised. Different authors also used different groupings. Breugel et al. (2007) included both short-lived and long-lived pioneer species in the group 'pioneers,' while Kennard (2002) classified species into groups according to Pinard et al. (1999), who divided short-lived and long-lived pioneers into two groups and also differentiated partially shade-tolerant species from shade-tolerant species.

4.3.5 Time to Reach Similar Species Composition as the Old-Growth Forest

Previous sections discussed how species composition changes along the succession trajectory. Studies demonstrated a different rate of change, where over time species composition gradually becomes more similar to that of local old-growth forests (Peña-Claros 2003).

Williams-Linera et al. (2011) showed in a study in the tropical dry forest of Mexico that the proportion of shade-tolerant species in 6 year-old fallows and an old-growth forest is similar, with shade-tolerant species comprising 54 % of species measured in the fallow and 67 % of species measured in the forest. Intermediate species accounted for 28 % of those measured in the fallows and 22 % of those measured in the forest, while pioneer species only accounted for 18 % of the total number of species in the fallows and 11 % in the forest (Williams-Linera et al. 2011). In the tropical moist forest of the Bolivian Amazon, however, a chronosequence study of 2–40 years-old fallows by Peña-Claros (2003) showed that the time needed to reach a species composition similar to that found in the old-growth forest is different between forest layers. Correspondence

Table 4.3 Summary of the number of years needed to reach similar species composition as the old-growth forest, from different studies

Place	Forest type	Years to reach similar species composition with old-growth forest	Source
Mexico	Lowland tropical rain forest and semi-deciduous forests	5 years to reach 60 %	Breugel et al. (2007)
Bolivia	Tropical moist forest	More than 100 years for the canopy layer; 40 years for the lower layers	Peña-Claros (2003)
Mexico	Tropical dry forest	6 years	Williams-Linera et al. (2011)

analyses were performed to estimate how fallow age affects species composition in the canopy, sub-canopy and understory layers of the forest. The results showed that lower layers old-growth forest species levels faster than the higher canopy layers. While lower layers need about 40 years of fallow to attain old-growth forest species composition, it might take about 100 years for canopy layers to reach a similar composition (Peña-Claros 2003). These authors suggested that the relatively rapid rate in the lower layers of the forest may be due to the continuous recruitment of long-lived pioneer and shade-tolerant species to the understory by the time of early succession (see also Uhl 1987). In contrast, the slower rate in the canopy layer could be explained by the dominance of pioneer and long-lived pioneer species recruited in early fallow stages (Uhl et al. 1981; Uhl 1987) with a high growth rate (Swaine and Hall 1983; Bazzaz 1991).

Table 4.3 shows the different times needed for species to attain a composition similar to that of the old-growth forest. Great variation could be observed. The number of years needed to attain a species composition similar to that of the old-growth forest ranged from 6 to 100 years. Even in the same forest (for example, the tropical moist forest of Bolivia), the canopy and lower layers also demonstrated great variation. In the following section, we will discuss the high variability of changes in species composition, and will examine various models that attempt to understand species composition in this context.

4.3.6 High Variability of Species Composition

Different models that predict changes in species composition during succession have been demonstrated in studies in the literature. A few studies concluded that the successional pattern they observed is congruent with the 'initial floristic composition' (IFC) model and the reductionist theory. This is the case, for example, in a chronosequence study conducted in the tropical dry forest of Mexico (Williams-Linera et al. 2011), in a study conducted in the tropical moist forest of the Bolivian Amazon (Peña-Claros 2003), and in a study conducted in the tropical rainforest of Eastern Madagascar by Klanderud et al. (2010). In these studies, forest species and pioneers established together during the early successional stage.

On the other hand, chronosequence studies conducted in the tropical moist forest of the upper Rio Negro region (Saldarriaga et al. 1988) and the evergreen broad-leaved forest in north-western Vietnam (Tran et al. 2010a, b, 2011) concluded that pioneer species recruit first at the beginning of succession. These results appear to be more congruent with the holistic approach than with the reductionist theory. Uhl (1987) concluded that both successional patterns are observed in the tropics.

Chazdon et al. (2007) suggested in their review that both the 'rely floristics' model and the 'initial floristic composition' model are not fully able to describe data from tropical forests. Kennard (2002) demonstrated in a study in Bolivia that long-lived pioneers dominated during the first year of succession (Fig. 4.5). The IFC model (Box 1), described by Finegan (1996) as the general descriptive model for neotropical succession, predicts that the long-lived pioneers (phase 3 in Box 1) replace the short-lived pioneers (phase 2 in Box 1) between 10 and 30 years. However, Kennard (2002) found that species from all functional groups colonized at the beginning of succession in the forest of Bolivia. This clearly does not fit the strict definition of the IFC model. Likewise, Breugel et al. (2007) concluded in a chronosequence study of the tropical rain forests and semi-deciduous forests in Mexico, that although both pioneers and forest species established together at early successional stages, their results are not entirely congruent with the IFC hypothesis. Breugel et al. (2007) found that all selected plots (fallows ranging from 1 to 5 years in age) already had closed canopies by the time of the first census. When a second census was carried out after 18 months, the recruitment by pioneer species was, on average, 59 % (Breugel et al. 2007). In one of the one-year-old plots, the pioneer *Cecropia peltata* was the most abundant species appearing on the fallows, with a relative abundance of 53 % (Breugel et al. 2007). However, this species had a high mortality rate, and accounted for approximately 86 % of the total stem mortality. At the same time, another pioneer species, *Trichospermum mexicanum* (which initially had a relative abundance of 31 %) replaced the former as the most abundant species, due to a much lower mortality rate (less than 5 % of total stem mortality) (Breugel et al. 2007). The IFC model predicts that pioneers will be replaced and that recruitment will decline after canopy closure (Box 1; Finegan 1984, 1996). The high average level of pioneer recruitment seems incongruent with the model, but the authors also stated that not all of the plots surveyed showed this high rate of pioneer recruitment. Thus, Breugel et al. (2007) concluded that the results of their study partly support the IFC hypothesis (see Fig. 4.8a), as the shade-tolerant fraction of the recruitment group was relatively higher, demonstrating a trend of gradual replacement of pioneers by shade-tolerant species. However, because of the high variation between fallows during the early successional stage, the authors suggested that there is a need for a new approach to explain the differences observed among sites during succession.

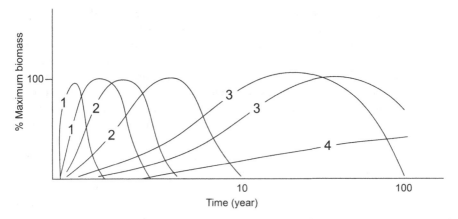

Fig. 4.9 General description of neotropical secondary succession. *Source* Finegan (1996)

Box 1 General Description of Neotropical Secondary Successions

According to Finegan (1996), during the first 100 years of succession in the neotropical lowland rain forest, three different recruitment phases can be indentified (Fig. 4.9). During these phases, canopy height, tree density, and basal area might reach similar levels to those observed in the mature forests in less than 30 years, provided that there is no soil degradation, and seed sources are near.

The first phase is dominated by herbs, shrubs and climbers, and is represented by a 1 in Fig. 4.9. Seedlings of the fast-growing, short-lived pioneer trees quickly emerge, and form a canopy within three years (2 in the figure). Vegetation of the first phase gradually disappear, as these species cannot survive under shade. The short-lived pioneer trees can survive between 10 and 30 years, and form a nearly even-aged population.

The long-lived pioneers (3 in figure 4.9) probably always colonize the field in the first few years of fallow, which means that their population is also approximately even aged. Phase three can last for approximately 75–150 years, which is the life span of the long-lived pioneer species. During the process of competition among species, colonization of the more shade-tolerant species (4 in the figure) is probably a continuous process. There are two main factors that are assumed to underlie changes in succession: (1) the intercorrelated differences in plant growth rate, tolerance of shade, age and size at maturity among species, most of which colonize the field shortly after abandonment, and (2) competition among individuals of those species.

Saldarriaga et al. (1988) concluded in their chronosequence study of the upper Rio Negro region that even with the same age, species composition may differ within stands. For example, the authors found that species like *Protium* sp., *Caryocar gracile* Wittm and *Cedrelinga* sp. could dominate in one stand, but could then be completely absent from another stand with the same age of fallow (Saldarriaga et al. 1988). Furthermore, Breugel et al. (2007) suggested that succession should not always be considered as a process in which the changes in functional groups are continuous, as the findings of their study showed that high mortality during the early stage of fallow could provide opportunities for newly recruited pioneers to take hold of the plots. Thus, we can conclude that species composition is highly variable, which makes it difficult to explain the patterns of species turnover and replacement. Therefore, species composition alone should not be used as an indicator of forest maturity, stability, or structural and functional characteristics (Breugel et al. 2007; Saldarriaga et al. 1988).

4.4 Discussion and Conclusions

A number of researchers (Finegan 1996; Corlett 1992; Clark 1996; Guariguata et al. 1997; Chazdon et al. 2007; Guariguata and Ostertag 2001) have suggested that more time was needed for fallows to attain a species composition similar to that of the old-growth forest because long-lived pioneer species tend to dominate the fallows for a very long time. A case study in Nigeria illustrates this trend. Researchers found that species richness, tree density and species diversity displayed considerable similarities with levels reported in the old-growth forest within 10 years; in contrast, the species composition of 10-years-old fallows showed considerable differences from that of the old-growth forest (Aweto 1981). Peña-Claros (2003) also found that a longer period of time was necessary for fallows to reach a species composition similar to that of the old-growth forest, whereas species diversity and species richness in the tropical moist forest of Bolivia recovered relatively quickly. Peña-Claros concluded that between 10 and 30 years were needed for total species richness and diversity to attain old-growth forest levels, while more than 100 years in the canopy layer and 40 years in the lower layers were required to regenerate a species composition similar to that of the old-growth forest. Similar results were found by Kammesheidt (1998) in the tropical moist forest of Paraguay. An analysis using the Sorensen index demonstrated that species composition between younger and older fallows had the highest similarity (Soerensen index: 0.89 in the tree stratum) when compared to younger fallows with old-growth forest (Soerensen index: 0.28 in the tree stratum), and older fallows with old-growth forest (Soerensen index: 0.44 in the tree stratum). Kammesheidt (1998) pointed out that these findings indicate that a long time is needed for fallows to regenerate to old-growth-forest levels of species composition, especially when comparing other forest structures, such as stem density, species richness and diversity (Kammesheidt 1998).

On the other hand, Guariguata and Ostertag (2001) emphasized the importance of recovering the same species composition of the uncut forest on fallows, in order to determine the function of the forest (e.g. nutrient cycling, net primary productivity). The authors point out that according to experiments done by Ewel et al. (1991) succession takes place without a great difference at the ecosystem level, regardless of species composition or richness. Ewel et al. (1991) observed the changes of three succession communities in Costa Rica throughout a period of 5 years following slash-and-burn. The three succession communities they observed were: (1) plants that colonized the site naturally (succession), (2) species that were deliberately planted, which had physiognomic characteristics similar to the natural succession vegetation (imitation of succession), and (3) more than 20 exotic species added to the field (enriched succession) (Ewel et al. 1991). Results showed that there were no significant differences in the nutrient retention properties of the three different communities. For example, the amount of soil organic matter on top soils was 13.6 % under 'succession,' 12.3 % under 'enriched succession,' and 12.7 % under 'imitation of succession' (Ewel et al. 1991). The amount of nitrogen, extractable phosphorus and extractable sulphur also showed similar trends. After 5 years, the amount of soil nitrogen was 0.73 % in 'succession' communities, 0.75 % in 'enriched succession' communities and 0.74 % in 'imitation of succession' communities (Ewel et al. 1991). For phosphorus and sulphur, the amounts under 'succession' were 992 and 1,281 mg/kg, under 'enriched succession' were 1,024 and 1,267 mg/kg, and under 'imitation of succession' were 992 and 1,354 mg/kg, respectively (Ewel et al. 1991). Thus, Guariguata and Ostertag (2001) concluded that these experiments indicated that 'the ability of species to occupy a site and use resources may be as important as the composition and diversity of species that colonize an area.' (p. 197).

In conclusion, changes in species composition along succession are highly variable. In different successional stages, species composition changes, and different species dominate. A small number of studies found that both pioneer and forest species recruit in early fallow, yet some studies suggested that pioneer species recruit first. Researches also found that different times are needed for species composition to reach levels that are comparable to old-growth forest. As suggested by Breugel et al. (2007), new approaches are needed to explain the great variability of species composition among sites, as neither the 'reductionist' nor the 'holistic' approaches seem to be able to fully predict and explain the regeneration pathway of species composition along the successional trajectory. The next chapter will review the local factors that contribute to the alteration of species composition, which need to be addressed to better understand the patterns of regeneration of the secondary forests.

References

Aweto AO (1981) Secondary succession and soil fertility restoration in south-western Nigeria I, succession. J Ecol 69:601–607

Bazzaz FA (1991) Regeneration of tropical forests: physiological responses of pioneer and secondary species. In Gómez-Pompa A, Whitmore TC, Hadley M (eds) Rain forest regeneration and management. Man and the biosphere series, vol 6. UNESCO, Paris, pp 91–118

Breugel MV, Bongers F, Martínez-Ramos M (2007) Species dynamics during early secondary forest succession: recruitment, mortality and species turnover. Biotropica 35:610–619

Chazdon RL, Letcher SG, van Breugel M, Martínez-Ramos M, Bongers F, Finegan B (2007) Rates of change in tree communities of secondary Neotropical forests following major disturbances. Philos Trans Royal Soc B 362:273–289

Clark D (1996) Abolishing virginity. J Trop Ecol 12:735–739

Corlett RT (1992) The ecological transformation of Singapore, 1819–1990. J Biogeogr 19:411–420

Delang CO (2007) Ecological succession of usable plants in an eleven-year fallow cycle in Northern Lao P. D. R. Ethnobot Res Appl 5:331–350

Egler FE (1954) Vegetation science concepts: I. Initial floristic composition—a factor in old-field vegetation development. Vegetatio 4:412–417

El-Sheikh MA (2005) Plant succession on abandoned fields after 25 years of shifting cultivation in Assuit, Egypt. J Arid Environ 61:461–481

Ewel JJ, Mazzarino MJ, Berish CW (1991) Tropical soil fertility changes under monocultures and successional communities of different structure. Ecol Appl 1:289–302

FAO (2000) FAO forestry, country profile: Madagascar [Online]. www.fao.org/forestry/site/6473/en/mdg, verified 27 June 2005

FAO (2003). Web site: http://www.fao.org/docrep/007/j2578f/J2578F21.htm

Finegan B (1984) Forest succession. Nature 312:109–114

Finegan B (1996) Pattern and process in neotroipcal secondary rain forests: the first 100 years of succession. Trends Ecol Evol 11:119–124

Fukushima M, Kanzaki M, Hla MT, Minn Y (2007) Recovery process of fallow vegetation in the traditional Karen swidden cultivation system in the Bago mountain range, Myanmar. Southeast Asian Stud 45(3):317–333

Gómez-Pompa A, Vazquez-Yanes C (1981) Successional studies of a rain forest in Mexico. In: West DC, Shugart HH, and Botkin, DB (eds) Forest succession, concepts and application. Springer-Verlag, New York, pp. 246–266

Guariguata MR, Ostertag R (2001) Neotropical secondary forest succession: changes in structural and functional characteristics. For Ecol Manag 148:185–206

Guariguata MR, Chazdon RL, Denslow JS, Dupuy JM, Anderson L (1997) Structure and floristics of secondary and old-growth forest stands in lowland costa rica. Plant Ecol 132:107–120

Halwagy R (1963) Studies on the succession of vegetation on some islands and sand banks in the Nile near Khartonm, Sudan. Vegetatio 6:218–234

Horn HS (1976) Succession. In: May RM (ed) Theoretical ecology: principles and applications. Blackwell Scientific Publications, Oxford, pp 187–204

Humbert H (1927) La destruction d'une flore insulaire par le feu: principaux aspects de la végétation à Madagascar: Document photographique et notices, vol 79. Mémoires de l'Académie Malgache

Kammesheidt L (1998) The role of tree sprouts in the restoration of stand structure and species diversity in tropical moist forest after slash-and-burn agriculture in Eastern Paraguay. Plant Ecol 139:155–165

Kennard DK (2002) Secondary forest succession in a tropical dry forest: patterns of development across a 50-year chronosequence in lowland Bolivia. J Trop Ecol 18:53–66

Klanderud K, Mbolatiana HAH, Vololomboahangy MN, Radimbison MA, Roger E, Totland Ø, Rajeriarison C (2010) Recovery of plant species richness and composition after slash-and-burn agriculture in a tropical rainforest in Madagascar. Biodivers Conserv 19:187–204

Knight DH (1975) A phytosociological analysis of species rich tropical forest on Barro Colorado Island, Panama. Ecol Monogr 45:259–284

Lebrija-Trejos E, Bongers F, Pérez-García Meave J (2008) Successional change and resilience of a very dry tropical deciduous forest following shifting agriculture. Biotropica 40:422–431

Lowry PP II, Schatz GE, Phillipson PB (1997) The classification of natural and anthropogenic vegetation in Madagascar. In: Goodman SM, Patterson BD (eds) Natural change and human impact in Madagascar. Smithsonian Institution Press, Washington, DC, pp 93–123

Metzger JP (2003) Effects of slash-and-burn fallow periods on landscape structure. Environ Conserv 30(4):325–333

Miller PM, Kauffman JB (1998) Effects of slash and burn agriculture on species abundance and composition of a tropical deciduous forest. For Ecol Manag 103:191–201

Mitja D, de Souza Miranda I, Velasquez E, Lavelle P (2008) Plant species richness and floristic composition changes along a rice-pasture sequence in subsistence farms of Brazilian Amazon, influence on the fallows biodiversity (Benfica, State of Pará). Agric Ecosyst Environ 124:72–84

Molina Colón S, Lugo AE (2006) Recovery of a subtropical dry forest after abandonment of different land uses. Biotropica 38:354–364

Neba NE (2009) Cropping systems and post-cultivation vegetation successions: agro-ecosystems in Ndop Cameroon. J Hum Ecol 27(1):27–33

Osunkoya OO, Aah JE, Hopkins MS, Graham AW (1994) Influence of seed size and seedling ecological attributes in shade-tolerance of rain-forest tree species in northern Queensland. J Ecol 82:149–163

Peña-Claros M (2003) Changes in forest structure and species composition during secondary forest succession in the Bolivian Amazon. Biotropica 35(4):450–461

Pinard MA, Putz FE, Rumiz D, Guzman R, Jardim A (1999) Ecological characterization of tree species for guiding forest management decisions in seasonally dry forests in Lomerio, Bolivia. For Ecol Manag 113:201–213

Raharimalala O, Buttler A, Ramohavelo CD, Razanaka S, Sorg JP, Gobat JM (2010) Soil-vegetation patterns in secondary slash and burn successions in Central Menabe, Madagascar. Agric Ecosyst Environ 139:150–158

Romero-Duque LP, Jaramillo VJ, Pérez-Jiménez A (2007) Structure and diversity of secondary tropical dry forests in Mexico, differing in their prior land-use history. For Ecol Manag 253:38–47

Rouw A (1993) Regeneration by sprouting in slash and burn rice cultivation, Taï rain forest, Côte d'Ivoire. J Trop Ecol 9:387–408

Saldarriaga JG, West DC, Tharp ML, Uhl C (1988) Long-term chronosequence of forest succession in the upper Rio Negro of Colombia and Venezuela. J Ecol 76:938–958

Schmidt-Vogt D (1998) Defining degradation: the impacts of swidden on forests in Northern Thailand. Mt Res Dev 18(2):135–149

Styger E, Rakotondramasy HM, Pfeffer MJ, Fernandes ECM, Bates DM (2007) Influence of slash-and-burn farming practices on fallow succession and land degradation in the rainforest region of Madagascar. Agric Ecosys Environ 119(3–4):257–269

Swaine MD, Hall JB (1983) Early succession on cleared forest land in Ghana. J Ecol 71:601–627

Toledo M, Salick J (2006) Secondary succession and indigenous management in semideciduous forest fallows of the Amazon Basin. Biotropica 38(2):161–170

Tran P, Marincioni F, Shaw R (2010a) Catastrophic flood and forest cover change in the Huong river basin, central Viet Nam: A gap between common perceptions and facts. J Environ Manag 91(11):2186–2200

Tran VD, Osawa A, Nguyen TT (2010b) Recovery process of a mountain forest after shifting cultivation in Northwestern Vietnam. For Ecol Manag 259:1650–1659

Tran VD, Osawa A, Nguyen TT, Nguyen BV, Bui TH, Cam QK, Le TT, Diep XT (2011) Population changes of early successional forest species after shifting cultivation in Northwestern Vietnam. New For 41:247–262

Uhl C (1987) Factors controlling succession following slash-and-burn agriculture in Amazonia. J Ecol 75:377–407

Uhl C, Clark K, Clark H, Murphy P (1981) Early plant succession after cutting and burning in the upper Rio Negro region of the Amazon basin. J Ecol 69:631–649

Von Schulthess L (1990) Inventaire de l'évolution des formations secondaires commebase pour leur conversion en forêts de production à l'exemple de Morondava sur la côte ouest de Madagascar, ETH Zurich

Walters MB, Reich PB (1996) Are shade tolerance survival and growth linked? Lowlight and nitrogen effects on hardwood seedlings. Ecology 77:841–853

Walters MB, Kruger EL, Reich PB (1993) Growth, biomass distribution and CO2 exchange of northern hard- wood seedlings in high and low light: relationships with successional status and shade tolerance. Oecologia 94:7–16

Williams-Linera G, Alvarez-Aquino C, Hernández-Ascención E, María T (2011) Early successional sites and the recovery of vegetation structure and tree species of the tropical dry forest in Veracruz, Mexico. New For 42:131–148

Chapter 5
Factors Contributing to Differences in Forest Recovery Rates

Abstract This chapter reviews the factors that affect forest regeneration, apart from fallow duration: the number of slash-and-burn cycles, fallow length, the type of soil, and the type of forest. Most research agrees that these factors influence forest regeneration processes: fewer slash-and-burn cycles, lower shifting cultivation intensity, and more fertile soils leads to greater biomass accumulation, higher plant density, higher tree height and larger basal area. However, there are still some uncertainties, in particular in how they affect the regeneration of fallows in the long-term.

Keywords Slash-and-burn cycles · Length of fallow period · Shifting cultivation intensity · Soil fertility · Soil texture · Forest type · Human disturbance

5.1 Introduction

Previous chapters reported that studies found considerable differences in the changes in species richness and diversity, plant density, vegetation abundance and species composition in fallows during succession. These differences implied that forest regeneration is determined by a complex set of factors besides fallow duration. As such, Chazdon et al. (2007) suggested that the rate of forest regeneration is affected by the complex interactions of different anthropogenic and natural factors, such as land use history, sites conditions, and regional species composition (see also Pickett et al. 1987). Similarly, a number of researchers (including Williams-Linera and Lorea 2009; Miller and Kauffman 1998; Molina Colon and Lugo 2006; Romero-Duque et al. 2007) argue that human disturbance and other environmental factors, such as past land use intensity and history, are likely to have an impact on the succession process. This chapter reviews the main factors that are expected to influence forest regeneration rates: the number of

C. O. Delang and W. M. Li, *Ecological Succession on Fallowed Shifting Cultivation Fields*, SpringerBriefs in Ecology, DOI: 10.1007/978-94-007-5821-6_5,
© The Author(s) 2013

slash-and-burn cycles (the number of times the fields were burned), fallow length, the type of soil, and the type of forest.

5.2 Number of Slash-and-Burn Cycles

An example from eastern Madagascar demonstrated the impact that land use intensity has on forest succession, since species abundances were found to be correlated with the number of slash-and-burn cycles (Klanderud et al. 2010). A series of Generalized Linear Model (GLM) multiple regressions showed that an increase in slash-and-burn cycles lead to a decrease in (1) species richness and abundance of tree seedlings, (2) species richness of saplings, (3) abundance of shrubs. At the same time, GLM multiple regressions showed that an increase in slash-and-burn cycles lead to an increase in (1) the abundance of herbs and (2) the abundance of adult trees (Klanderud et al. 2010). The reasons for these changes, which accompanied an increase in slash-and-burn intensity, were hypothesized to be related to the influence of fire and the lack of time and opportunity for tree seedlings to germinate.

According to Tran et al. (2011), soil seed banks would decline with continued burning (see also Uhl et al. 1981) As soil organic matter and litter layers are removed, seeds stored below the surface would be removed as well (see also Howlett and Davidson 2003). Guariguata and Ostertag (2001) also argued in their review that if slash-and-burn intensity increases, soil would be disturbed more frequently by fire, which would pose a threat to soil seed banks, while removing stumps and roots, which would also limit the ability of plants to resprout (see also Ewel et al. 1981; Uhl et al. 1981). For example, Uhl et al. (1981) looked at how the cutting and burning of forest would affect the size of soil seed bank and sprouts roots, by selecting three sites in the forest of the upper Rio Negro Region of the Amazon Basin, two of which were cut in September 1976 and burned in December 1976, and one of which was cut in August 1978 and burned in November 1978. After burning, Uhl et al. (1981) reported a decline in the size of the soil seed bank: the mean number of seedlings per sample in the undisturbed mature forest was 30.1, while on the burned fallow it was only 6.3. The difference was significant, which indicated that fire posed a threat to the seed bank. On the other hand, sprout roots were also depleted after burning, as out of 121 sprouted shoots found in the site before the fire, only one was able to survive the fire (Uhl et al. 1981). Moreover, the authors pointed out that after 4 months of burning, only a few new shoots could be sprouted (0.6 m^2), which indicated that fire would destroy resprouting ability within the fallow (Uhl et al. 1981).

Ewel et al. (1981) examined secondary forests with 8 to 9-year-old fallows in the tropical premontane wet forest of Costa Rica to understand the effect of fire on the changes in soil seed bank. Soil seed bank was measured by the number of seeds geminated after slashing and burning, and the results showed a decline in the number of seeds that germinated after the fire when compared with undisturbed forests (Fig. 5.1). The number of seeds that germinated in the forest was about

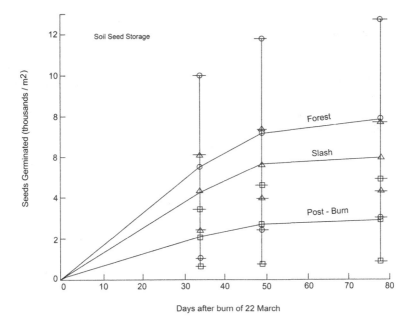

Fig. 5.1 Seeds germinated from soil sampled from the forest, the 11-week-old slash, and after the burn. Center symbols of each type on the vertical bars indicate the mean. Upper and lower symbols mark ± 1 SD. Circles refer to forest, triangles to slash, and squares to post-burn. *Source* Ewel et al. 1981

8,000 seeds/m^2; in contrast, after burning the number of seedlings found was <3,000 seeds/m^2 (Ewel et al. 1981). The authors concluded that 52 % of the soil seed bank that was present beneath the mulch before the burn was killed by the fire. Similar results were also obtained by Tucker et al. (1998) in their comparison of Altamira and Bragantina (Brazil). These authors concluded that repeated burning would kill the seed bank and that only a few species (those resistant to fire) would survive. The authors argued that the slower growth of plant density, stand height and basal area in Bragantina might be due to the longer slash-and-burn history of Bragantina, wherein shifting cultivation had been practiced for more than 100 years (Denich 1991). This was in stark comparison to the mere 25 years of cultivation history of Altamira (Moran et al. 1994). With repeated cycles of forest cutting and burning, Tucker et al. (1998) pointed out that the soil seed bank would be diminished, although the authors did not provide data to quantify the reduction in the number of seeds presented in their sites. The authors suggested that seed source in Bragantina was further limited by the cutting of all primary forest in the region, thus resulting in a retarded forest succession (Tucker et al. 1998). In contrast, Altamira experienced a shorter slash-and-burn history, allowing it to benefit from the colonization of seeds from the primary forest (Tucker et al. 1998).

Besides the depletion of soil seed banks, other researchers found that fire may have a negative impact on the ability of the forest to resprout. Figure 5.2 (Rouw

1993) shows the effect of fire on limiting the growth of resprouting species: the number of resprouting species under normal burning is less than that under burning with low intensity (slightly burn). However, this difference gradually declines as fallows age. Milleville et al. (2000) also found that repeated burning leads to soil degradation, resulting in a grass savannah fallow because of increased soil erosion and compaction.

Sproutings is deemed to be important in forest regeneration as mentioned by a number of studies in the literature (Tran et al. 2010b; Rouw 1993; Vieira and Proctor 2007). After examining succession in Vietnam, Tran et al. (2010a) suggested that sprouting from stumps and roots was essential for forest regeneration, because it served as a major catalyst for plant recruitment at the beginning of succession (see also Rouw 1993; Vieira and Proctor 2007). In the lowland evergreen rain forests of the eastern Amazon, Vieira and Proctor (2007) investigated three sites with 5-, 10- and 20-year-old fallows with at least eight cropping cycles each. The importance of sprouting was shown in the early stage of succession. For small plants (≥ 1 m tall, <5 cm dbh), the mean density and number of species that regenerate from sprouts were 5.4 and 37, respectively, while the density of seeds was 4.6 and the number of species that originated from seeds was 36 in a 5-year-old fallow (Vieira and Proctor 2007). However, the relative importance of sprouts was found to decrease with fallow age. In 10-year-old fallows, the number of plants that originated from seeds outnumbered those regenerated from sprouts. In a 10-year-old plot, 58 plants originated from seeds and 19 plants originated from sprouts, while in 20-year-old fallows, 48 plants originated from seeds and 28 plants originated from sprouts (Vieira and Proctor 2007). However, sprouting was shown to be the principle form of regeneration for larger plants with dbh greater than five cm. Vieira and Proctor found that throughout all age classes, more plants originated from sprouts than seeds: in 5-year-old fallows, 18 plants originated from sprouts and 4 from seeds; in 10-year-old fallows, 25 plants originated form sprouts and 13 from seeds; and in 20-year-old fallows, 31 plants originated from sprouts and 13 from seeds (Vieira and Proctor 2007). These results underscore the importance of sprouting in early succession, while in later stages plant establishment from seeds increases and sprouting becomes less significant (Vieira and Proctor 2007).

In common with Vieira and Proctor (2007), the permanent plot study conducted in the rain forest of Côte d'Ivoire (Rouw 1993) following shifting cultivation with a cropping period of about 2 years concluded that the number of resprouting species decreased as succession proceeded (Fig. 5.2). The study also revealed that 70 % of the total number of resprouting species is comprised of primary forest species. Kammesheidt (1998) conducted a chronosequence study in the tropical moist forest of Paraguay and concluded that the proportion of sprouts decreased with fallow age in the sapling stratum, with values declining from 59.5 % of total number of stems in 2- to 5-year-old fallows to 32.9 % in 10 and 15 year-old fallows. Additionally, no pioneers that originated from seeds were found in the sapling stratum in 4- and 5-year-old fallows. However, in the tree stratum, Kammescheidt observed a different trend. The proportion of stems that originated from sprouts showed a similar value in both young and old fallows, accounting for

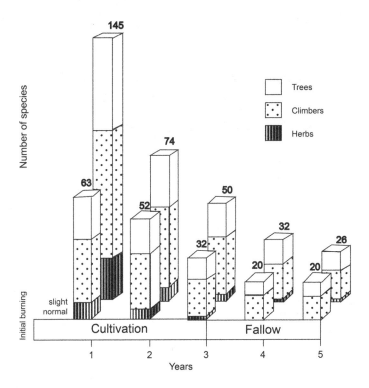

Fig. 5.2 The number of tree, climber, and herbaceous species resprouting during years of cultivation and fallow in field 1, Taï forest. Observations were made in permanent plots (6 × 9 m²) where the initial burning had been normal, and where the initial burning had been slight (6 × 9 m²). *Source* Rouw 1993

21 and 19.6 % of the total number of stems, respectively (Kammesheidt 1998). Despite the difference found in the sapling and tree stratum, the author concluded that sprouts were essential to forest regeneration in the early stage of succession. For instance, species such as *Inga uruguensis* and *Trema micrantha* had an Importance Value Index of 21.8 and 20.1, respectively, which means that they were both commonly found in early fallow stages that regenerate through sprouting (Kammesheidt 1998). However, when succession proceeded, the importance of resprouting lessened as the percentage of sprouts in basal area decreased with fallow age, and the percentage of plants that originated from seeds increased (Fig. 5.3; Kammesheidt 1998). A longer chronosequence was observed by Saldarriaga et al. (1988), who examined fallows ranging in age from 9 to 80 years old. The number of sprouts ranged from 22 to 4,516/ha, with the higher number in younger fallows (Fig. 5.4). Thus, sproutings were more prevalent in the early stage than they were in the older ones, including in the old-growth forest.

Tran et al. (2010a), Vieira and Proctor (2007), Rouw (1993) and Kammesheidt (1998) all pointed out the importance of resprouting in succession, especially in

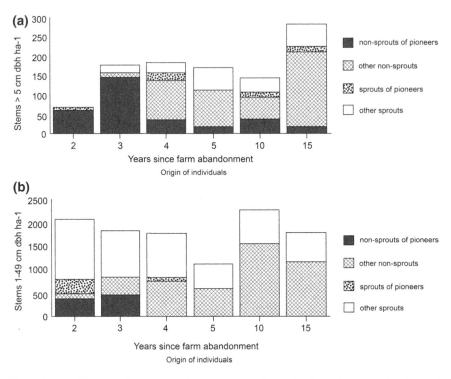

Fig. 5.3 a, b Number of sprouts and non-sprouts of woody plants in successional stands following slash-and-burn agriculture in Paraguay. *Source* Kammesheidt 1998

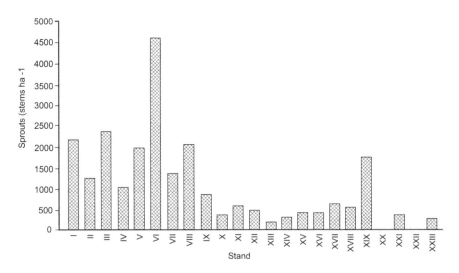

Fig. 5.4 Number of sprouts in successional and mature forest stands in the upper Rio Negro region of Colombia and Venezuela. *Source* Saldarriaga et al. 1988

Table 5.1 Comparison of the rates of recovery of basal area, tree species richness and canopy height reported by other studies of forest succession in the tropics and subtropics. For each index, the age by which secondary stands reach 75 % of the mature forest stand, as well as the mature stand values, are reported

Forest type	Basal area (m² ha⁻¹)		Canopy height		Tree species richness		Source
	Age at 75 % (y)	Mature (m² ha⁻¹)	Age at 75 % (y)	Mature	Age at 75 % (y)	Mature (no.)	
Montane	>>30	60.3	>>30	36	15	20	Kappelle et al. 1996
Lower montane	>35	62.8	11	21	>35	105[a]	Kuzee et al. 1994
Subtropical wet	20	33.8	20	24	>21	37	Brown and Lugo 1990a
Subtropical wet					25	20	Aide et al. 1995
Tropical moist	>80	34.8	>35	25–35	60	67	Saldarriage et al. 1986
Tropical most	>80	35.6			5	66	Saldarriaga et al. 1988
Tropical wet	17	33			>17	70	Guariguata et al. 1997
Tropical dry			>10	10.4	10	25	Aweto 1981a, b
Tropical dry	23	25	23–40	25	5	37	Kennard 2002

>> signifies that the mature stand value was more than twice that of the oldest secondary stand reported
[a] Total species richness
Source Kennard 2002

the early stage. The high proportion of sprouting in the early successional stage might allow an earlier and quicker growth of plants so that taller and larger crowns could form at the beginning of succession (Miller and Kauffmann 1998). Kennard (2002) found higher rates of regeneration of basal area, canopy height and tree species richness as compared with other studies (Table 5.1). This study pointed out that the faster growth rate of forest structure might be due to the high rate of sprouting in the forest of Bolivia, as sprouts grow faster than seedlings in general, and forests with a high proportion of species that originate from sprouts are deemed to have greater resilience to disturbance (see also Corlett 1981; Ewel 1980; Janzen 1975; Nyerges 1989). Thus, sprouting appears to be significant for plant regeneration during succession, although its importance may decline in later stages of succession. Despite these studies, it is still unclear exactly how sprouting affects the succession process in later stages, and what its impact on the forest recovery rate is. More research covering longer periods of succession is needed in order to understand the effect of sprouting on forest regeneration.

Using Canonical Correspondence Analysis (CCA), Klanderud et al. (2010) showed that both the number of slash-and-burn cycles and the age of the fallow were associated with the growth of tree seedlings, although the intensity of slash-and-burn activity was found to contribute more to tree seedlings establishment than these other factors. On the other hand, species richness, species composition and species abundance (of tree saplings) were affected more by the number of years since the field had been abandoned than by the number of slash-and-burn cycles. This indicated that the number of times the field had been burned is the determining factor for tree seedlings' germination and growth, while fallow duration determines if the seedlings can grow into saplings or not Klanderud et al.(2010). As a result, the authors suggested that the time needed for recovery could be reduced if the number of slash-and-burn cycles decreased and the fallow duration increased. The ability of tree seedlings to germinate is essential for tree regeneration on fallows as suggested by Tran et al. (2010a), who pointed out that stem density in the seedling stratum and in early fallow stages showed obvious variation. This implies that the seedling stage is a sensitive time for tree growth.

In short, research suggested that slash-and-burn intensity affects tree seedlings' establishment, and that the number and richness of species would decline, as concluded by Klanderud et al. (2010) in their study on the forests of Eastern Madagascar. However, despite fire depleting the soil seed bank and destroying the sprouting roots, it is still unclear how the succession process affects plots after cultivation in the long-run. More detailed studies are needed to examine the effects of the number of slash-and-burn cycles on the regeneration rate of fallows, on forest function such net primary productivity, on nutrient recycling in the soil (Guariguata and Ostertag 2001), and on the recovery of the forest structure in general.

5.3 Length of Fallow Period

Shifting cultivation practices are found in different places in the world, and may be classified into two types: 'establish swidden cultivation' (also called 'rotation shifting cultivation' or 'secondary forest shifting cultivation') and 'pioneer swidden cultivation' (also called 'primary forest shifting cultivation') (Fukushimas et al. 2008; Walker 1975; Grandstaff 1980; Fukushima et al. 2007). 'Establish swiddening' is practiced by the Karen, Khmu, Htin and Lawa in the hilly area of northern Thailand, and the Karen in Burma (Fukushima et al. 2008; Tran et al. 2011; Fukushima et al. 2007). 'Pioneer swiddening' is practiced by Hmong, Lisu, Lahu, Akha and Yao ethnic groups in northern Thailand and Hmong people in Vietnam (Fukushima et al. 2008; Tran et al. 2011). 'Established swiddening' is less intensive than 'pioneer swiddening,' as fields are left fallow for 6–12 years after 1 year of cropping. In contrast, 'pioneer swiddening' involves clearing the land and farming it for several years, until yields drop (Keen 1978; Fukushima et al. 2008; Tran et al. 2011). However, the number of years the field can be farmed before yields drop excessively depends also on the crop being cultivated. For example, Keen (1978) found that the Hmong in Thailand could farm the same field with poppy for 10 to 15 consecutive years, but with rice and maize only for 2 or 3 years.

Researchers suggested that the intensity of shifting cultivation practices would impact the succession process and influence the growth rate of the forest structure (Uhl 1987; Tran et al. 2010b; Fukushima et al. 2008; Fukushima et al. 2007; Kennard 2002). Uhl (1987) conducted a study in San Carlos in the upper Rio Negro region, where sites with different cropping duration were examined for forest structure and species composition. After 5 years of fallow, species richness,

Table 5.2 A comparison of vegetation structure and biomass for sites in the vicinity of San Carlos de Rio Negro, Venezuela, subjected to moderate (Sites 2 and 3) and high (Site 4) levels of disturbance

	Site 2	Site 3	Site 4
Type of land use	Forest cut, burned and farmed for 3 years		Forest cut, burned and farmed intensively for 6 years
Intensity of farming	Moderate (traditional farming)		High (prolonged farming)
Years since abandonment	5	5	5
Soil	Oxisol	Oxisol	Ultisol
Canopy height (m)	11	9	4
Basal area (m^2/ha)	7.4	8.1	2.8
Biomass (t/ha)			
Stem	30.0	26.5	7.1
Leaf	3.8	3.7	2.8
Total	33.8	30.2	9.9
Mean number of tree species ≥ 2 m tall/100 m^2 plot	9	9	4

Source Uhl 1987

basal area, canopy height and biomass were 2–4 times higher in sites which had been farmed for 3 years than in sites with 6 years of uninterrupted farming (Table 5.2). Results indicated that high intensive farming would retard forest succession. Likewise, species richness was found to be lower under intensive shifting cultivation practices in the evergreen broad-leaved forest of north-western Vietnam (Tran et al. 2010b). The study was carried out in the mountain area, where 80 % of the population was Hmong. Here, they were practicing high intensity shifting cultivation wherein land was cropped for 4–5 years (Tran et al. 2010b). The authors concluded that under intensive cultivation, a slow recovery rate was observed. They found 35 species in 26-year-old fallows compared with 72 species in the old-growth forests (thus, 26-year-old fallows reached approximately 49 % of the species richness of the old-growth forest). On the other hand, 20- to 30-year-old fallows in the less intensive fields farmed in northern Thailand (where farmers conducted 1 year of cropping with 6–10 years of fallow), had approximately 67 % of the 103 species found in the old-growth forest. This is a much faster rate of growth than that in Vietnam (Fukushima et al. 2008).

Researchers also found that the less intensive way of forest clearing by the Karen would lead to a faster rate of forest regeneration (Fukushima et al. 2007; Fukushima et al. 2008; Kanjunt and Oberhauser 1994; Schmidt-Vogt 1998, 1999). Fukushima et al. (2007) studied forest regeneration in a Karen area in the Bago mountain range in Burma, and observed that stumps were cut at one meter above the ground. On the other hand, researchers were unable to find any remains of trees and stumps after burning in land cleared by the Hmong in Thailand (Keen 1978). Fukushima et al. (2007) suggested that the stumps and roots that remained after slash-and-burn could facilitate recovery at later stages as sprouting is possible from the stumps and roots, so trees would be able to recover more quickly. In this study, the number of years needed to reach an equivalent amount of aboveground biomass to that of the old-growth forest is less than reported in the other studies (Table 2.1). In northern Thailand, the Karen used similar methods that lead to faster recovery (Fukushima et al. 2008). The authors pointed out that the faster forest recovery was due to the less intensive way of cutting, together with the short period of cultivation and long fallow period (see also Schmidt-Vogt 1998, 1999). Schmidt-Vogt (1998) carried out a study comparing the different swidden culti-vation methods practiced by ethnic groups in the evergreen montane forest of northern Thailand in three different villages—the Lawa in Ban Tun, the Karen in Ban Huai Sai and the Akha in Ban Aze. The author observed that in Ban Tun, the Lawa practiced a low-intensity swidden cultivation, wherein large trees were left on the field during forest clearing, while small trees with dbh <12 cm were cut. The density of the relicts ranged from 66 to 383 individuals/ha, and the fields were cultivated for 1 year for rice, and then usually left fallow for a period of 17 years (Schmidt-Vogt 1998). On the other hand, the Karen in the village of Ban Huai Sai used a swiddening method similar to that of the Lawa, but farmlands were cropped for a longer period of between 2 and 10 years, and fewer relicts were left. The author found that the 10 year-old stand was floristically and structurally similar to the 12 year-old fallow in Ban Tun, but the stand in Ban Huai Sai only contained about half of the number of species found in Ban Tun (31 species were recorded in Ban Tun but only 16 were found in Ban Huai Sai stand). Furthermore, fallows in Ban Huai Sai were dominated by only two species—*Quercus aliena* and *Myrica esculenta*—while the stand in Ban Tun had high levels of species diversity.

The Akha of Ban Aze migrated to Thailand in 1981, and settled on land previously cultivated by the Hmong, who had practiced intensive swidden culti-vation (Schmidt-Vogt 1998). According to Schmidt-Vogt (1998), the Hmong cultivation duration was between 6 and 8 years, but could vary from 1 to 8 years. He suggested that with a different cultivation duration, the succession process would vary, with species like *Chromolaena odorata* becoming dominant in fields that have been farmed for a long period of time, as stumps and roots were removed. Small grass species might then become dominant on the fields, as *Chromolaena odorata* is unable to survive burning. Thus, grassland might develop on fallows that had been cropped for a long period of time (Schmidt-Vogt 1998). On the other hand, the author hypothesized that for land that had been cropped for a short period of time (1 or 2 years), fallow would develop differently, as larger grasses such as bamboo, would thrive. Schmidt-Vogt (1998) concluded in his study that fallows in Ban Aze could be regarded as degraded when compared to stands in Ban Tun and Ban Huai Sai. This is because when the author first started the field work, fallows in Ban Aze were 17 years old, and the number of species found was similar to the field in Ban Huai Sai. However, Ban Aze fallows had a lower tree height and wider spacing between trees, which the author suggested was an indicator of delayed forest development (Schmidt-Vogt 1998). The less intensive swidden cultivation practice by the Lawa and Karen thus seems to facilitate forest regeneration (Schmidt-Vogt 1998).

Nyerges (1989) carried out a study in the tropical deciduous forest of Sierra Leone, and demonstrated a different succession pathway on fields with higher versus lower cultivation intensity. The study area was located in Kilimi, a Susu community, and the author suggested that the way of cultivation practiced by the Susu maintained a low level of damage to stumps and roots. Nyerges (1989) described their way of swidden cultivation as 'minimal cultivation,' as land is usually cropped for 1 year and left fallow for 15–30 years, after which trees are cut at one m aboveground. Land was not cleared at once but in stages, whereby small stems were cut first, so that the canopy was not lost at once, and the soil could be protected under the canopy (Nyerges 1989). The author observed some of the large trees with diameters of 27–300 cm would be left on the field; for example, in one of the 30-year-old fallows, 29 trees remained after forest clearing, or about 15 relicts/ha. After burning, the author reported that 18 % of the land cover remained as stumps or branches. Under this low-intensity farming practice, Nyerges (1989) found a high rate of coppicing: in the 1-year-old fallow, 84 % of saplings grew from coppice roots; in the 3-year-old fallow, 57 % grew from coppice roots, in the 7-year-old fallow, 73 % grew from coppice roots, and in the 25-year-old fallow, 61 % grew from coppice roots. This demonstrates that coppicing dominates in the total basal area after just 3 years of fallow (Fig. 5.5). Saplings that originated from coppice shoots accounted for two thirds of all the sapling stems during the succession process. Furthermore, the plants that grow shortly after harvest originate from coppicing from trees left before cropping rather than new seedlings. When succession proceeds, the majority of species that are found on the sites are coppice species. These include *Dialium guineense*, *Monodora tenuifolia*, *Bridelia micrantha*, *Ochthocosmus africanus*, *Anthonotha macrophylla*, *Amphimas pterocarpoides*, *Anthostema senegalense*, and *Anisophyllea laurina*. Under this 'minimal cultivation' practice, the author stated that the succession process takes place immediately after land is abandoned, and that the forest canopy is fully recovered after 10–30 years of fallow.

However, cultivation practice in the Kilimi area displayed some variations, as higher-intensity practices could be indentified (Nyerges 1989). It was found that 87.5 % of the total cultivated area was farmlands that exploited fallows. Some of these farmlands experienced a longer cropping period, with 15 % cropped for 2 years and 2.3 % farmed for 3 years (Nyerges 1989). Nyerges (1989) found that when comparing the succession process on fallowed sites farmed only with rice, or for 1 year with rice and a second year with groundnuts, fallows farmed for 1 year with rice had a greater number of plants established by coppicing. On the other hand, sites that farmed for 2 years (1 year of rice and 1 year of groundnuts) were found to have a high rate of grass invasion (in particular, *Chasmopodium caudatum* and *Andropogon* spp.) (Nyerges 1989). The author suggested that grass would invade more easily when farming practices failed to allow a canopy to develop. This invasion by grass species eventually leads to the death of stumps. Thus, Nyerges suggested that farmland cultivated for one or two additional years might restrain coppicing ability, which would cause a slower rate of canopy growth, and subsequent invasion by pioneers. However, Nyerges (1989) did not give further information as to the frequency or amount of the invaded grass and pioneer species,

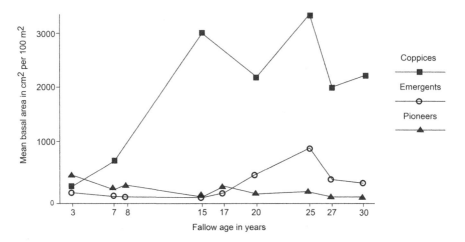

Fig. 5.5 Coppices, emergents, and pioneers in forest fallows aged 3–30 years. *Source* Nyerges 1989

and a more detailed study should be conducted to prove the validity of the reduced rate of canopy growth.

Apart from using a longer cropping duration, people in the Kilimi area also practiced shorter fallow periods. 2.2 % of the farmlands were cleared from fallows that were 6 years old or younger (Nyerges 1989). Results showed that the coppice ability of the fallows that had previously been farmed for 1 year on fallow sites that were <6 years old was lower than the sites that were farmed on old fallows (30 years or older). The rate of coppicing immediately after land abandonment was 84 % on land farmed on young fallows and 91 % on land farmed on old fallows (Nyerges 1989). This same study found that the percentage of coppicing species of young fallow and old fallow for one-year-old rice crops was 17 and 24 %, respectively, which showed that coppicing ability declines when fallow length is shortened.

In short, Nyerges (1989) concluded that the coppicing is significant in decid-uous forests as coppice species could grow in a short period of time after land was abandoned, and before the seedlings of other species establish and germinate. This has important implications for forest regeneration. Thus, the increase in shifting cultivation intensity (with longer cropping periods and shorter fallow durations) might retard the recovery rate during the succession process. As discussed pre-viously, sprouting was found to be essential in the early stage of forest succession. In Burma (Fukushima et al. 2007), for example, researchers showed that a higher sprouting rate can lead to a faster rate of forest structure growth along the suc-cession process. More research on the long term effect of sprouting and shifting cultivation intensity would be beneficial for the local communities whose liveli-hoods depend on shifting cultivation.

As suggested by Mitja et al. (2008), the intensity of shifting cultivation practice not only impacts the changes of forest structure during succession, but also influenced species composition. Their study, conducted in the tropical rain forest of the Brazilian Amazon, concluded that if the intensity of forest clearing is low, the species composition in the early stage will agree with the reductionist approach and the IFC model, where both pioneer and forest species grow together at the beginning of succession. On the other hand, if the clearing intensity is high, pioneers will colonize first, resulting in the successional characteristics described by the holistic model (Mitja et al. 2008). Research by Klanderud et al. (2010) in the forest of eastern Madagascar agreed with the findings of Mitja et al. (2008), who found that sites with low intensity clearing before cultivation could protect plant roots and soil seed banks, which might lead to both pioneers and forest species being recruited together in the early stage of succession.

However, as mentioned in the previous section, some studies concerning succession following shifting cultivation found that species composition followed trends that more closely fit the holistic approach than the IFC model (e.g. Saldarriaga et al. 1988; Tran et al. 2010b, 2011). The research in Vietnam by (Tran et al. 2010b, 2011) was conducted in forests settled by Hmong, who practiced high intensity shifting cultivation. This may be the reason that species composition was found at those sites to more closely fit the holistic approach. On the other hand, Saldarriaga et al. (1988) did not mention the slash-and-burn method practiced by the people in the upper Rio Negro region; therefore, we are not able to conclude if the intensity of shifting cultivation lead to the successional process more closely fitting the holistic approach. Thus, it is difficult to predict the pathway of species composition development and how the degree of shifting cultivation intensity affects species composition. Currently, this is hard to measure in the field, where numerous variables may impact the results obtained by researchers.

In short, the intensity of shifting cultivation might impact forest regeneration during the successional process. Although studies suggest that a longer duration of farming and higher intensity of clearing might have a negative impact on forest regeneration, there is no clear consensus on this issue. How the reduction in sprouting and coppicing in fallows affects succession in the long run is a question that future work still needs to address.

5.4 Type of Soil

Another set of factors found to have an influence on succession processes is soil properties. Chazdon et al. (2007) stated that differences in soil properties would lead to local variations in vegetation structure and composition (see also Herrera and Finegan 1997; Mendez-Bahena 1999; Finegan and Delgado 2000); however, there is some disagreement in the literature, as we now discuss.

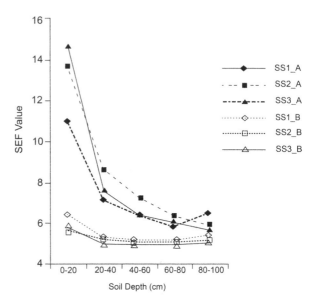

Fig. 5.6 Comparison of SEF values in Altamira (**a**) and Bragantina (**b**) SS1 is initial secondary succession, SS2 is intermediate secondary succession, SS3 is advanced secondary succession. *Source* Lu et al. (2002)

5.4.1 Faster Forest Regeneration in Soil with Higher Fertility and Better Structure

Some of the studies in the literature concluded that fallows would have a faster regeneration time if the soil is more fertile and has a better structure. Lu et al. (2002) found a correlation between soil nutrient levels and soil physical structure with biomass growth rate after land abandonment. The soil evaluation factor (SEF), which is used to evaluate soil fertility, was examined in two areas: Altamira (Alfisols) and Bragantina (Oxisols and Ultisols). In Altamira, the SEF was much higher than that measured in Bragantina from early through advanced stages of succession (Fig. 5.6), which indicated that Alfisols in Altamira are more fertile than the Ultisols and Oxisols indentified in Bragantina. Table 5.3 also shows that soil nutrients like Ca, Mg, K, N and organic matter in the soil of Altamira had higher concentrations than they did in Bragantina. Furthermore, clay contents appeared to be higher in Altamira soils (clay and silt content was approximately 80 %). Soils in Bragantina had a higher level of coarse and fine sand (at least 70 %), which would have caused a more rapid loss of nutrients; thus, soil fertility was lower (Lu et al. 2002). This indicates that Alfisols had a better physical structure than Ultisols and Oxisols (Lu et al. 2002).

The average biomass growth rates of the two sites were different as well. In Altamira, the biomass growth rate increased when fallows aged and was higher then that in Bragantina throughout the succession process. In contrast, the biomass growth rate dropped in Bragantina when succession proceeded (Fig. 5.7). Thus, vegetation recovered more quickly in Altamira than in Bragantina during succession. The authors stated that the reason behind this may have been the higher nutrient

Table 5.3 Comparison of soil properties in different succession stages

Type	Depth	pH	Ca	Mg	K	Al	N	OM	Coarse	Fine	Silt	Clay
Altamira (Alfisols)												
SS1	0–20	5.15	2.15	0.6	0.07	0.15	0.19	2.19	15	19	12	55
	20–40	5.4	1	0.5	0.03	0.15	0.15	1.44	10	12	14	65
	40–60	5.35	0.8	0.5	0.01	0.15	0.1	1.11	7	9	16	68
	60–80	5.45	0.7	0.45	0.02	0.2	0.1	0.8	8	11	11	70
	80–100	5.55	0.6	0.65	0.01	0.25	0.09	1.35	5	9	16	70
SS2	0–20	4.6	2.2	0.55	0.07	1.15	0.19	3.49	14	14	17	56
	20–40	4.9	1.7	0.4	0.02	1	0.11	2	14	12	14	61
	40–60	4.95	1.75	0.15	0.02	0.85	0.09	1.37	12	11	15	63
	60–80	5.1	1.25	0.3	0.01	0.9	0.07	1.06	11	10	14	66
	80–100	5.25	1.3	0	0.01	0.75	0.06	0.84	10	11	14	66
SS3	0–20	4.95	2.9	0.55	0.06	0.25	0.22	2.85	10	11	25	55
	20–40	4.9	1	0.7	0.04	0.25	0.16	1.59	8	9	21	63
	40–60	4.95	0.85	0.35	0.02	0.2	0.11	1.2	7	9	23	61
	60–80	5	0.7	0.45	0.01	0.15	0.1	0.95	7	9	19	65
	80–100	5.3	0.65	0.55	0.01	0.05	0.1	0.58	7	9	21	64

(continued)

Table 5.3 (continued)

Type	Depth	pH	Ca	Mg	K	Al	N	OM	Coarse	Fine	Silt	Clay
Bragantina (Oxisols and Ultisols)												
SS1	0–20	5.16	0.76	0.34	0.02	0.48	0.05	1.55	39	37	14	10
	20–40	4.92	0.36	0.22	0.02	0.72	0.05	0.75	34	36	11	19
	40–60	4.98	0.28	0.2	0.01	0.72	0.04	0.82	32	32	14	22
	60–80	5.1	0.3	0.18	0.01	0.7	0.04	0.85	33	31	12	24
	80–100	5.06	0.32	0.18	0.01	0.68	0.03	1.03	32	34	10	24
SS2	0–20	4.67	0.33	0.25	0.03	0.8	0.06	1.79	42	34	13	11
	20–40	4.75	0.25	0.15	0.03	0.83	0.04	1.4	40	30	13	17
	40–60	4.8	0.27	0.18	0.02	0.87	0.04	0.75	35	31	12	22
	60–80	4.92	0.28	0.16	0.02	0.82	0.04	0.47	32	33	12	23
	80–100	4.9	0.3	0.2	0.02	0.72	0.03	0.74	35	29	11	25
SS3	0–20	4.93	0.47	0.23	0.03	0.7	0.06	1.53	48	31	12	9
	20–40	5.036	0.2	0.13	0.01	0.9	0.05	1.26	42	29	11	18
	40–60	5.03	0.13	0.07	0.01	0.7	0.04	0.95	41	28	11	20
	60–80	5.1	0.15	0.1	0.01	0.7	0.04	0.79	39	28	11	21
	80–100	5.27	0.13	0.1	0.01	0.43	0.04	0.86	38	29	11	22

Note The unit for soil depth is cm; the unit for Ca, Mg, K, and Al is Meq 100 g^{-1}; the unit for N, OM, and soil structure (coarse, fine, silt, and clay) is percent
Source Lu et al. (2002), Table 5.1

Fig. 5.7 Average biomass
growth rates (kg/m²/year) in
Altamira and Bragantina.
Source Lu et al. (2002)

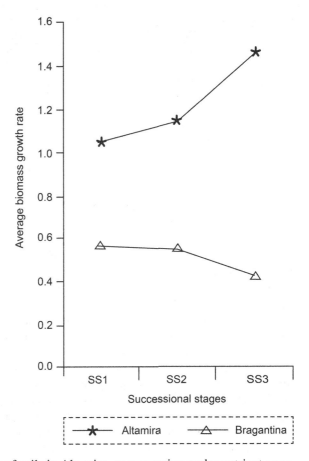

content and better structure of soils in Altamira, as succession under nutrient-poor,
coarse, and sandy soils could be slow and difficult. At the same time, because of a
faster accumulation of biomass in Altamira, soil fertility increased as fallows aged
(Lu et al. 2002). SEF in the soil surface layer showed a large increase from SS1 to
SS2 and SS3 in Altamira (Fig. 5.6). Thus, according to the authors, an interrelationship between vegetation and soil nutrient cycling was well maintained in
Altamira. On the contrary, SEF in the soil surface layer dropped in Bragantina when
fallows aged, and the soil nutrients tended to concentrate along the soil profile
(Fig. 5.6). Lu et al. (2002) suggested that biomass accumulation was not fast enough
to offset the loss of nutrients due to leaching. Furthermore, the study further
examined the significance of soil properties with respect to the rate of vegetation
growth by using regression coefficients (Lu et al. 2002). Biomass accumulation rate
was significantly influenced by the amount of soil nutrients such as Ca, K and Al, and
the fine sand content (Table 5.4). As a result, Lu et al. (2002) showed that soil with a
higher level of fertility (as measured by SEF) and better structure would facilitate
biomass growth rate and vice versa, while at the same time, the amount of vegetation
grown would also affect soil nutrient accumulation.

Table 5.4 Relationships between soil properties and average yearly biomass growth rate

Study area	Depth	R	Significant variables
Altamira	0–20	0.788[a]	N %, Silt, Clay, Fine, Mg, Al, OM %, K
(Alfisols)	20–40	0.791[a]	OM %, Clay, Fine, Mg, N %, Al, Ca
	40–60	0.630	Fine, N %, Ca
	60–80	0.595	Fine, Ca
	80–100	0.588	No variables
Bragantina	0–20	0.828[a]	Ca, Mg, Al, K
(Ultisols and	20–40	0.798[a]	K, Fine, OM %
Oxisols)	40–60	0.689[a]	Al, OM %, Fine
	60–80	0.850[a]	Al, OM %, Fine, Ca
	80–100	0.882[a]	Al, N %, Fine, OM %

Note [a] Indicates that regression coefficient is significant at 0.05 level
Source Lu et al. 2002, Table 5.2

Another study carried out in the Altamira and Bragantina regions of the Amazonian forest in Brazil by Tucker et al. (1998) examined 15 sites in Altamira and 16 sites in Bragantina where fields usually cropped for 1–2 years, with a chronosequence of 6–16 years. Tucker et al. (1998) concluded that fallows with more fertile soils had a faster recovery rate in their forest structures, including plant density, stand height and basal area. Throughout the chronosequence, the average density of saplings was double in Bragantina than in Altamira. In the first 2 years of succession, Altamira had about 8,600 saplings/ha, which declined when fallows aged. On the other hand, while Bragantina had some 6,000 saplings/ha when reaching the 3rd year, the number continued to increase until year 10 of fallow (Tucker et al. 1998). The authors also found a higher number of trees in Altamira than in Bragantina: 773 trees/ha were found in the 7-year-old fallows of Altamira, but only 407 trees/ha were found in the 8-year-old fallows in Bragantina. Thus, more time was needed for saplings in Bragantina to establish and grow, which correspondingly slowed down the succession process (Tucker et al. 1998).

The changes in tree height of the two sites were also found to be distinct during succession. Altamira demonstrated a higher stand height throughout the succession process (Tucker et al. 1998). The maximum tree height in the first 5 years was recorded as 10 m in Altamira and 3.75 m in Bragantina. Between years 6 and 10, the maximum tree height increased to 22.5 m in Altamira and 13 m in Bragantina (Tucker et al. 1998). When succession proceeded in the older age class of 14–16 years, the maximum height increased to 35 m in Altamira, but it still remained 13 m in Bragantina (Tucker et al. 1998). The maximum total height serves as an indicator of the growth potential of the fallows, and the authors concluded, based on this measure, that stands in Bragantina needed twice the amount of time to reach the same height as Altamira, which showed a slower rate of growth. On the other hand, total and tree basal area were reported to be higher in Altamira than Bragantina throughout the succession period. Tucker et al. (1998) found that in Bragantina, the total basal area was 2.7 m^2/ha in the 4-year-old fallows, while the basal area of 0- to 5-year-old stands in Altamira was 7 m^2/ha. For tree basal area, the average of 6- to 10-year-old fallows in Bragantina was 3.83 m^2/ha, only one fifth of that recorded in Altamira, where it was 15.07 m^2/ha. When succession advanced, the 20-year-old fallow in Bragantina had a tree basal area of approximately 9.71 m^2/ha, while in Altamira the tree basal area of a 7-year-old fallow was 11.36 m^2/ha, already higher than in Bragantina (Tucker et al.1998).

The differences in the changes in forest structure during succession in the two sites were due to the variations in soil fertility and texture among the sites as previously discussed (see above). The soil fertility index (SFI) was used to evaluate the influence of soil fertility on the growth of fallows (Moran et al. 1996). The results of an ANOVA test showed that tree height as well as variations in forest structure between the two sites were significantly affected by soil fertility: stand height showed an increase of 0.16 m when the index increased by one unit (Tucker et al. 1998). Moreover, the authors pointed out that soil texture affected forest regeneration. Soils in Altamira were mainly clayey, while soils in Bragantina were sandy. The authors suggested that sandy soils are less fertile as the ability to hold nutrients, minerals and water is lower (Brady 1974), which led to slower forest regeneration in Bragantina (Tucker et al. 1998). On the other hand, the authors suggested that the high clay content of Altamira soil leads to the ability of plots in this site to retain nutrients and minerals. As a result, nutrients can stay in the soil long enough for plants to uptake them, and at the same time water can also be captured to avoid drought (Jordan 1985). The forests of Altamira had a faster growth rate because of nutrient-rich Alfisols, in comparison to the forests of the Bragantina sites which were dominated by nutrient-poor Oxisols and Spodosols (Tucker et al. 1998).

Similarly to the research of Lu et al. (2002), a study conducted in the upper Rio Negro region by Saldarriaga et al. (1988) found that soil fertility strongly influenced the biomass growth rate. Saldarriaga et al. (1988) pointed out that about 189 years would be needed for the above-ground biomass to recover to a level equivalent to that of the old-growth forest, which appeared to be longer than other areas in the tropics. For instance, in central Burma, about 40 years are needed to attain an amount of above-ground biomass comparable to that of the old-growth forest (Fukushimat et al. 2007) and approximately 65–120 years would be needed in the southern Yucatan Peninsular Region in Mexico (Read and Lawrence 2003). The amount of K, Ca and Mg found in the upper Rio Negro region were 10–130 times lower than in Costa Rica (Hardy 1961; Harcombe 1977). Thus, the slower rate of biomass recovery might be due to poor soils (Oxisols and Ultisols) in the upper Rio Negro region (Saldarriage et al. 1988).

In short, most researchers concluded that forests regenerate faster in plots with more fertile soils. However, not all studies agreed with this finding. Some found that soil properties had no influence on forest regeneration.

5.4.2 Soil Properties Have No Effect on Forest Structure

Despite some studies that reported that soil properties impact the rate of change in forest structure during succession, Raharimalala et al. (2010) concluded that differences in soil type does not lead to significant variations in species richness along the successional trajectory. However, the authors did report that species composition varied on different soils. Raharimalala et al. (2010) monitored a total of 61 fallows with different age classes on two types of soils—yellow and red—in the dry deciduous forest of south-western Madagascar. Red soils contained high amounts of sand or gravel and had clay contents between 5 and 10 %, which were poor in nutrients and humus layers. Yellow soils had a higher clay content of between 10 and 15 %, as well as a higher pH, and a normal permeability. These soils were thus richer than red soils (Raharimalala et al. 2010). After running an ANOVA, the authors reported that the richer yellow soils showed no higher species richness than fallows on red soils, although yellow soils reported a higher number of weeds (Raharimalala et al. 2010). However, species composition was found to be different in the two soil types. According to a redundancy analysis, results show that some of the species were associated with soil parameters. For example, species like Z. mays (zea), M. esculenta (manes) and Arachis sp. (asp) tended to grow in soil with a high pH (predominantly yellow soils); while H. contortus (heco), M. arcuata (Musa), S. grevioides (sidg), M. ambatoana (mamb), C. lamii (jacp) and F. madagascariensis (fgre) were all associated with acidic soils (predominantly red soils). On the other hand, A. bernieri (aber) and Cryptostegia grandiflora (chgr) were correlated with carbon- and nitrogen-rich soils, while C. bosseri (cybos), Breweria sp. (bsp) and Triumfetta sp. (trium) were associated with nutrient poor soils (Raharimalala et al. 2010).

5.4.3 Impact of Shortened Fallow Duration on Soil Fertility

As mentioned above, it is generally accepted that when fallows age, tree species will eventually dominate the landscape. Therefore, it logically follows that if fallow length is shortened, grass species will dominate and be unable to reach tree fallow (Neba 2009; Styger et al. 2007), which would lead to a more homogeneous landscape (Metzger 2003). Indeed, research in the semi-deciduous tropical forests of Mexico found that the frequency of shrubs and herbs increases with shorter fallow (Dalle and Blois 2006). After examining 26 sites with fallow ages ranging from 5 to 40 years, Dalle and Blois found from their redundancy analysis that the species that have the strongest relation with short fallows (<10 years) are shrubs like Viguiera dentate, Melanthera aspera, Desmodium incanum and, to a lesser extent, Centrosema schotti. On the other hand, longer fallows had more tree species than shrub species (for example, they reported tree species such as Vitex gaumeri and Lysiloma latisiliquum). Similar results were found in Cameroon. Neba (2009) reported that a short fallow of 1 to 5 years is typical with slash-and-burn cultivation in the rain forest of Cameroon, and leads to a savannization process. A total of 130 fields with different fallow durations from five villages were randomly selected for the study. Neba (2009) observed that during the first year of fallow, the land was characterized by weeds. In the second and 3rd years, grasses such as *Imperata cylindrical* and *Hyparrhenia* spp. invaded and continued to dominate into the 4th year of fallow. The researchers reported that 5 year-old fallows were rare, but that taller *Hyparrhenia* grasses were the most abundant on those fields. Neba (2009) suggested that the shrub savanna landscape is a plagioclimax where anthropogenic disturbances prevent the fallows from developing towards a climatic climax. The limitation of tree growth on short fallows might be due to the great competition with weeds on the field (Metzger 2003). Metzger examined fallows of various ages in the north-eastern Brazilian Amazon and reported that when fallow duration was reduced, the spatial dominance of weeds and other species of the early successional stage are favoured, and that competition between weed seeds and pioneer tree seeds increases. Thus, as concluded by Metzger (2003), the rate of tree-seedling establishment and the process of regeneration would be adversely affected, and the conditions of germination for forest species would be limited.

Neba (2009) concluded that with the short fallow duration, land would be depleted, as the natural nutrient recycling by a mature forest is needed to maintain land quality. Farmers in Madagascar also stressed that the presence of woody species on fallows is essential for maintaining crop yields; therefore, the land should not be cultivated again before it reaches the *Savoka* stage (taller shrub fallow). Otherwise, woody species could not grow again, which would lead to land degradation and the development of an herbaceous fallow (Styger et al. 2007). Other researchers came to the same conclusions. Styger (2004) pointed out that herbaceous fallows are limited in their ability to accumulate biomass and to restore soil fertility, while Uhl (1987) stated that tree growth plays an important role in determining the recovery rate of the land, as the presence of pioneer tree species in the early stage of fallow could provide the suitable microenvironment necessary for woody species to establish and to improve soil nutrients. The author adopted the results from Uhl et al. (1982), where the authors examined three sites in the Rio Negro region of Venezuela after 1 year of fallow. A higher percentage of soil organic matter was found in the sites with trees in comparison with sites harbouring only bare soil (3.24 and 2.42 % respectively). Total Phosphorus (P) and total Nitrogen (N) also revealed a higher content in tree-bearing sites (166 mg/ 100 g for total N, and 18 mg/100 g for total P on site with trees; while 132 mg/ 100 g for total N, and 23 mg/100 g for total P on site with bare soils).

Thus, researchers generally agree that fallow length should be long enough for trees to regenerate, because without the chance for woody species to develop, soil fertility may not be restored, as trees are necessary to maintain soil nutrients. However, these studies have usually only focused on describing the changes in vegetation when fallow length is shortened, with a statement pointing out the importance of trees for maintaining soil fertility. These studies have neither directly investigated the relationship between soil-vegetation nor the mechanism

Table 5.5 Soil organic carbon, soil pH, and basal area of plots[a] in Tanzania

	Organic C (%)	pH	Basal area (m^2)[b]
Site 1 average	5.74[c]	4.4[c]	n/a
Natural forest	5.50[c]	5.0[c]	0.68
Field	4.39[c]	4.4[c]	-
Acacia mearnsii	7.41[c]	3.9[c]	0.62
Site 2 average	0.48[c]	5.6[c]	n/a
Thicket	0.47[d]	5.4[c]	-
Eucalyptus citriodora	0.44[d]	5.3[c]	0.19
Cassia siamea	0.62[d]	6.2[c]	0.46
Site 3 average	7.02[c]	6.8[c]	n/a
Natural forest	6.36[d]	6.8	1.8
Cupressus lusitannica	7.65[d]	6.8	1.27

Indicates not measured

n/a Indicates not applicable

[a] Plot values are means of 15 samples; site values are means of 45 samples

[b] Basal area of stems >cm dbh in 2 5 × 2 5 plots

[c] Significantly different from one another at .005 level

[d] Significantly different from one another at .05 level

Source Allen (1985)

of nutrient recycling by trees and how it could benefit soil fertility restoration (e.g. Neba 2009; Styger et al. 2007). However, one study by Allen (1985) pointed out that a positive relationship exists between organic carbon in the soil and the abundance of plants grown on the plots (Table 5.5). Furthermore, the amount of vegetation found on the fields represents the quantity of organic matter that goes back into the soil following land use.

Apart from soil nutrient depletion due to the shortening of fallow duration, the example from Madagascar shows that when fallow length is held constant along cropping cycles, soil fertility decreases, and fallow species eventually change from trees to grass (Styger et al. 2007). Styger et al. (2007) reported that fallow length should be longer for each cropping cycle, as crop yield showed a decline after the land was cropped a few times, even though fallow length is maintained. According to local farmers, the first fallow cycle is usually dominated by trees. However, starting from the second fallow cycle, the number of tree species dropped, and eventually, the land was occupied by shrubs from the third cycle. Farmers remarked that fallow duration should be at least 3 years for the first two cycles, 5 years for the third cycle and 8 years for the fourth cycle, to maintain soil fertility and rice yield.

Thus, it is suggested that when determining fallow length, the number of cropping cycles should be taken into account. However, while Styger et al. (2007) make this general point, the authors did not explain the causes behind this phenomenon or how soil fertility depletes when cropping progresses. There is very little research that examines this issue.

Despite the fact that a relationship appears to exist between shorter fallow and the reduction of trees, there is a lack of data on the relationship between soil nutrient depletion and tree cover. In addition, some researchers have reported that no relationship exists at all between soil fertility and fallow length. Two cases are now reviewed, one in which soil fertility increases with fallow age, and one in which it decreases.

5.4.3.1 Soil Fertility Increases with Fallow Age

A few studies in Southeast Asia (Nakano 1978; Toky and Ramakrishnan 1983; Bui 1990) and the Neotropics (Lamb 1980; Williams-Linera 1983; Werner 1984; Silver et al. 1996) have found that a positive relationship exists between soil fertility and fallow duration. For instance, a study was carried out on the south-western coast of Madagascar, where slash-and-burn agriculture is dominant (Raharimalala et al. 2010). Results showed that in the study area, plots with longer fallow lengths also contained more mature soil with a higher carbon content. As succession proceeded in this site, the total carbon and C/N ratios were very different among age classes. Total carbon content increased from 0.9 % on fallows with an age class of 1–5 years old, to 1.6 % in 11–20 years old plots. A permanent plot study conducted in the upper Rio Negro also concluded that total soil organic matter increased by 18 % after 5 years of fallow, where total Nitrogen (N), total Phosphorus (P) and other exchangeable chemical elements also reached levels seen in old-growth forests (Uhl 1987). For instance, after 5 years of fallow, the site that was farmed for 3 years was found to have a total N content of 151 mg/100 g, and total P content of 21 mg/100 g, while in the site that had been farmed for 6 years, the amount of total N and total P was 169 mg/100 g and 18 mg/100 g, respectively, after 5 years of fallow (Uhl 1987). Both sites showed very similar values to the 126 mg/100 g of total N and 18 mg/100 g of total P in the old-growth undisturbed forest (Uhl 1987). Another study was conducted by Lu et al. (2002) in the Amazon Basin, in which fallows of different stages following shifting cultivation and pasture were selected from two regions—Altamira and Bragantina. Fallows were classified into three successional stages: the initial stage (SS1), the intermediate stage (SS2) and the advanced secondary succession (SS3). In this study, Alfisols were found in Altamira, and Ultisoils and Oxisols were indentified in Bragantina (Lu et al. 2002). Lu et al. developed a SEF in order to evaluate soil fertility in different types of soil and successional stages:

$$SEF = [Ca + Mg + K - \log(1 + Al)]^{*}OM + 5$$

(Ca: Calcium, Mg: Magnesium, K: Potassium, Al: Aluminum)

A similar equation, the SFI, has been used in other studies (e.g. Moran et al. 2000a, b; Tucker et al. 1998) to examine soil fertility:

$$SFI = pH + OM + P + K + Ca + Mg - Al$$

(OM: Organic matter)

Lu et al. (2002) pointed out of the use of the SFI has some drawbacks. For instance, the units used to measure soil nutrients were different (Ca, Mg, K, and Al is Meq/100 g, while P is ppm, and OM is %). Also, Lu et al. (2002) argue that pH is not an independent value, as it is influenced by the amount of Ca, Mg and Al. The authors stated that the SEF they developed would be more comprehensive as the problem of the differences in units could be omitted (Ca, Mg, K, and Al is all Meq/100 g), and the constant 5 would avoid a negative value of SEF if a high amount of Al is in the soil.

From the study, they found that the SEF increases from initial to advanced succession in Altamira (Fig. 5.6), although the organic matter in the top soil (0–20 cm depth) slightly decreases from 3.49 to 2.85 % in the intermediate to advanced stage after an increase from 2.19 to 3.49 % in the initial to intermediate succession (Lu et al. 2002). In short, a positive relationship between soil fertility and fallow length has been demonstrated in several studies.

5.4.3.2 No Relationship Exists Between Soil Fertility and Fallow Age

There is also some evidence that no significant relationship exists between soil fertility and fallow age (Fukushima et al. 2007; Hughes et al. 1999). Fukushima et al. (2007) conducted a chronosequence study in the mixed deciduous forest of Burma where they examined the peak total carbon (TC) and total nitrogen (TN) levels in an 18 year-old fallow field following shifting cultivation. The authors used Scheffe's test and Spearman's test to see how significant the TC and TN levels between fallow age classes (including old-growth forest) were, and how strong the correlations were between TC and fallow length and TN and fallow length. Results showed that TC and TN levels varied among age classes with no significant difference between fallow stands and old-growth forest, and that neither TC nor TN levels were significantly correlated with fallow length. The authors suggested that not much organic matter is lost from the soil during cultivation because of the short cultivation duration at their sample site (only 1 year). This example shows that factors other than fallow length might contribute to the ability of soil nutrients to recover. Similar results were found by Hughes et al. (1999) and Reiners et al. (1994), both of whom conducted studies in Neotropical regions. Hughes et al. (1999) found that the amount of soil Carbon (C), Nitrogen (N) and Sulfur (S) did not change along a 50-year chronosequence in forests of Mexico, based on data demonstrating that total C, total N and total S levels in one-meter-depth soil did not correlate with fallow age. Total Carbon (C) ranged from 139 to 269 Mg/ha, total N ranged from 14000 to 24000 kg/ha, and total S ranged from and 2200 to 4500 kg/ha. On the other hand, net N-mineralization rates and net nitrification rates between fallows of 10–20 years old and primary forest were similar in forests of Costa Rica, where N-mineralization rates ranged from 12.61 to 17.64 mg/kg of soil for fallows, and 11.52–14.86 mg/kg of soil for old-growth forest (Reiners et al. 1994). Net nitrification rates for fallows were 12.80–17.39 mg/kg of soil, and for old-growth forest from 10.62 to 17.00 mg/kg of soil (Reiners et al. 1994).

In dry forests of Egypt, a chronosequence study showed that levels of organic matter in the soil increase in early fallow stages (from 1 to 3 years), but then decline from the 5th year onwards (El-Sheikh 2005). In this study, although woody species eventually dominated when fallows aged, accumulation of soil nutrients was not observed. The authors explained that this might be due to the rapid nutrient consumption by weeds in the early stages due to soil levels and leaching (see also: Odum 1960; Aweto 1981b; Odum et al. 1984; Lee 2002, pp. 138–155). However, the study did not provide data on the amount of organic soil matter; thus, El-Sheikh (2005) failed to examine the relationship between soil organic matter and fallow age.

5.5 Types of Forest

Kennard (2002) and Williams-Linera et al. (2011) stated that the succession process in tropical dry and humid forests is different, as dry forests show a faster rate of recovery in forest structure, species composition and diversity than humid forests. Although vegetation growth rate might be slower in dry versus wet forests, due to the lower height and simpler forest structure, dry forests were deemed to have a higher recovery rate and resilience (Kennard 2002; Ewel 1980; Murphy and Lugo 1986a, b). However, Quesada et al. (2009) pointed out that it is questionable whether the recovery rate of dry forests is faster than that of wet forests.

Quesada et al. (2009) pointed out that some researchers (e.g. Ewel 1977; Murphy and Lugo 1986a, b; Segura et al. 2003) suggested that the recovery rate of dry forests was faster than that of wet forests, as forest structure is relatively simpler and coppicing is predominant after disturbance. Yet, Quesada et al. (2009) remarked that there was insufficient evidence to support such claims, since the predominant method of plant regeneration in dry forests is through seeds, and case studies have shown that the genetic diversity of adult trees is high, which indicates that regeneration should be by seeds rather than coppicing (Hamrick and Loveless 1986; James et al. 1988). Thus, dry forests might not necessarily regenerate faster than wet forests (Quesada et al. 2009). Moreover, Chazdon et al. (2007) pointed out that in general, the accumulation rate of above-ground biomass in tropical forests increases with increasing levels of precipitation (see also Brown and Lugo 1982; Murphy and Lugo 1986a, b). However, a study conducted in the tropical dry forest of the Yucatan in Mexico found no relationship between total aboveground biomass (TAGB) and the regional precipitation gradient, as they instead observed small differences in the biomass within regions for the same stand age (Fig. 5.8; Read and Lawrence 2003). Although differences in precipitation within regions were observed, the small variability showed that precipitation is not a significant factor affecting TAGB, considering that greater regional biomass differences might be expected (Read and Lawrence 2003).

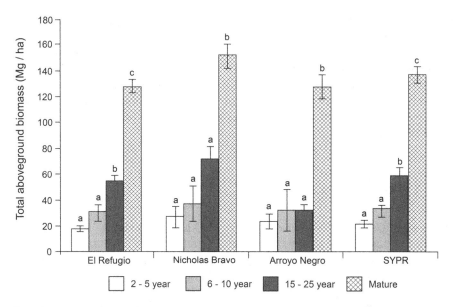

Fig. 5.8 Total aboveground biomass (TAG) by age class within each sampling region and in the southern Yucatan Peninsular Region overall. Results are of four ANOVAs (all $P < 0.001$) comparing TAG by age class within each region and in the SYPR overall. Bars depict means $\pm$ 1 SE, and those with different letters within each of the regions are significantly different at $\alpha = 0.05$ (Tukey post hoc test). *Source* Read and Lawrence (2003)

5.6 Discussion and Conclusions

In this chapter, we have discussed how four different factors—(1) slash-and-burn cycle, (2) shifting cultivation intensity, (3) soil condition and (4) forest types, impact the changes in forest structures and species composition. Most of the research on this topic has concluded that these factors influence forest regeneration processes.

In short, studies found that fewer slash-and-burn cycles, lower shifting cultivation intensity, and more fertile soils might lead to greater biomass accumulation, higher plant density, higher tree height and larger basal area. However, concerning soil fertility, other studies found no difference in species richness between more fertile and less fertile soils despite the fact that species composition showed a distinct pattern on different soils. How the ecological succession process following shifting cultivation is influenced by soil properties needs more research and attention, as not much research has been conducted to examine the effect of soil fertility and texture on vegetation changes (see Tucker et al. 1998 for a discussion). The interrelationship between soil and vegetation is an important issue when determining how long fallow cycles should be to maintain the productivity of the land.

In conclusions, studies could not determine how well tree species retain soil fertility on fallows. A study carried out in the forest of Burma suggested that the presence of other pioneer species in the early stage of fallows is essential for maintaining soil fertility at later stages (Fukushima et al. 2007). For example, Fukushima et al. (2007) reported a rapid invasion of the flowering shrub *Chromolaena odoratum* during the first year of fallow in Burma, which might explain why TC and TN levels in young fallows were not lower. The authors pointed out that Koutika et al. (2002) found that soil in Cameroon that contained this species was richer than soil in fields without it because this species helps prevent soil erosion. The presence of the species may protect organic soil matter, especially during the early regeneration stage. Thus, Fukushima et al. (2007) and Koutika et al. (2002) suggested that some herbaceous fallow vegetation—not only trees, as suggested by other studies—may contribute to soil nutrient restoration.

The restoration of forest soil fertility levels to pre-farming levels depends on different factors such as soil type (Guariguata and Ostertag 2001; Allen 1985; Weaver et al. 1987; Neill et al. 1997) and the intensity of land use before fallowing (Guariguata and Ostertag 2001; Buschbacher et al. 1988; Neill et al. 1997). These factors contribute to the ability of the fallow to repopulate soil nutrients. Thus, examining only fallow length is insufficient to understand the whole picture of soil restoration. For instance, after examining the soil properties of three different sites in north-eastern Tanzania, Allen (1985) suggested that soils with higher acidity tend to accumulate larger amounts of organic carbon (Table 5.5). Lu et al (2002) pointed out in their study that the physical structure of the soil would affect the ability of soil to accumulate nutrients. Their results show that soil nutrients, as well as organic matter, N, Ca, Mg, and K contents, are found to be higher in Altamira than in Bragantina, with soil in Altamira having higher clay and silt contents, and lower rates of coarse and fine sand (Table 5.3) Lu et al. (2002). Therefore, it is suggested that the clay content of the soil would have an impact on the soil's ability to retain nutrients (Lu et al. 2002).

With different opinions available in the literature, no conclusion can be drawn regarding whether dry forests have a faster recovery rate than wet forests. From our review, we could also not definitively conclude whether or not dry forests have a faster rate of regeneration than moist forests during succession. Moreover, the number of studies examining dry forests is rather small compared to those examining moist forests. As pointed out by Ruiz et al. (2005, p. 520), "tropical dry forests cover more area than humid forests (Whigham et al. 1990), but are among the least known tropical ecosystems (Rico-Gray et al. 1988; Martinez-Yrizar et al. 2000)". There is a need for more research on the ecological succession process of dry forests, so as to better understand the overall effect of climate on forest regeneration, and whether there are any predictable trends.

There is also a need for additional long-term studies to determine how much the number of slash-and-burn cycles, shifting cultivation intensity, soil condition and forest types impact forest structures, species richness and diversity, and species composition, as fallows age. In this chapter, we only examined four variables. However, there are likely more factors that contribute to the alteration of the successional process, such as distance to mature forests (Chazdon et al. 2007). To better understand how long a land should be left fallow in order to achieve an ideal crop yields in the next cultivation cycle, it is important to understand how the natural characteristics (e.g. soil condition, distance to forest, landscape effects) and anthropogenic influences of the land (e.g. land use history, slash-and-burn intensity) contribute to the changes in secondary forest succession patterns. More research is needed on the impact of environmental variables and anthropogenic disturbances on forest regeneration processes.

References

Aide M, Zimmerman K, Rosario R, Marcano H (1995) Forest recovery in abandoned cattle pastures along an elevational gradient in northeastern Puerto Rico. Biotropica 28:537–548

Allen JC (1985) Soil response to forest clearing in the United States and the tropics: geological and biological factors. Biotropica 17:15–27

Aweto AO (1981a) Secondary succession and soil fertility restoration in south-western Nigeria, I, succession. J Ecol 69:601–607

Aweto AO (1981b) Secondary succession and soil fertility restoration in south-western Nigeria. II, soil fertility restoration. J Ecol 69:609–614

Brady NC (1974) The nature and properties of soils. MacMillan Publishing Co., Inc., New York

Brown S, Lugo AE (1982) The storage and production of organic matter in tropical forests and their role in the global carbon cycle. Biotropica 14:161–187

Brown S, Lugo AE (1990a) Tropical secondary forests. J Trop Ecol 6:1–32

Brown S, Lugo AE (1990b) Effects of forest clearing and succession on the carbon and nitrogen content of soils in Puerto Rico. Plant Soil 124:53–64

Bui QT (1990) Some problems on slash-and-burn cultivation soil in Northwestern region of Vietnam and the direction of its utilization. PhD Dissertation, Hanoi

Buschbacher R, Uhl C, Serrão EAS (1988) Abandoned pastures in Eastern Amazonia. II, Nutrient stocks in the soil and vegetation. J Ecol 76:682–699

Chazdon RL, Letcher SG, van Breugel M, Martínez-Ramos M, Bongers F, Finegan B (2007) Rates of change in tree communities of secondary Neotropical forests following major disturbances. Philos Trans Royal Soc B 362:273–289

Corlett RT (1981) Plant succession on degraded land in Singapore. J Tropical For Sci 4:151–161

Dalle SP, De Blois S (2006) Shorter fallow cycles affect the availability of noncrop plant resources in a shifting cultivation system. Ecol Soc 11(2):2

Denich M (1991) Estudo da Importâcia de uma Vegetação Secundária Nova para o Incremento da Produtividade do Sistema de Produção na Amazonia Oriental Brasileira. Ph.D. Dissertation: Departamento de Ciências Agrárias da Universidade Georg August de Göttingen. EMBRA-PA/CPATU, Belém, Brasil

El-Sheikh MA (2005) Plant succession on abandoned fields after 25 years of shifting cultivation in Assuit, Egypt. J Arid Environ 61:461–481

Ewel J (1977) Differences between wet and dry successional tropical ecosystems. Geogr Ecol Tropical 1:103–117

Ewel JJ (1980) Tropical succession: manifold routes to maturity. Biotropica 12:2–9

Ewel J, Berish C, Brown B, Price N, Raich J (1981) Slash and burn impacts on a Costa Rican wet forest site. Ecology 62(3):816–829

Finegan B, Delgado D (2000) Structural and floristic heterogeneity in a 30-year-old Costa Rican rain forest restored, on pasture through natural secondary succession. Restor Ecol 8:380–393

Fukushima M, Kanzaki M, Hla MT, Minn Y (2007) Recovery process of fallow vegetation in the traditional Karen swidden cultivation system in the Bago Mountain Range. Myanmar, Southeast Asian Studies 45(3):317–333

Fukushima M, Kanzaki M, Hara M, Ohkubo T, Preechapanya P, Choocharoen C (2008) Secondary forest succession after the cessation of swidden cultivation in the montane forest area in Northern Thailand. For Ecol Manag 255:1994–2006

Grandstaff TB (1980) Shifting cultivation in northern Thailand, resource systems theory and methodology series, no. 3, United Nations University, Tokyo

Guariguata MR, Ostertag R (2001) Neotropical secondary forest succession: changes in structural and functional characteristics. For Ecol Manag 148:185–206

Guariguata MR, Chazdon RL, Denslow JS, Dupuy JM, Anderson L (1997) Structure and floristics of secondary and old-growth forest stands in lowland Costa Rica. Plant Ecol 132:107–120

Hamrick JL, Loveless MD (1986) Isozyme variation in tropical trees: procedures and preliminary results. Biotropica 18:201–207

Harcombe PA (1977) Nutrient accumulation by vegetation during the first year of recovery of a tropical forest ecosystem, recovery and restoration of damaged ecosystems. In: Cairns J, Dickson KL, Herricks EE (eds) University of Virgina Press, Charlottesville, pp 347–376

Hardy F (1961) The soils of the I.A.I.A.S area (Mimeo), Turrialba, Costa Rica

Herrera B, Finegan B (1997) Substrate conditions, foliar nutrients and the distributions of two canopy tree species in a Costa Rican secondary rain forest. Plant Soil 191:259–267

Howlett BE, Davidson DW (2003) Effects of seed availability, site conditions and herbivory on pioneer recruitment after logging in Sabah. Malaysia For Ecol Manag 184:369–383

Hughes RF, Kauffman JB, Jaramillo VJ (1999) Biomass, carbon, and nutrient dynamics of secondary forests in a humid tropical region of Mexico. Ecology 80(6):1892–1907

James T, Vege S, Aldrich P, Hamrick JL (1988) Mating systems of three tropical dry forest tree species source. Biotropica 30:587–594

Janzen D H (1975) Ecology of plants in the tropics. In: The institute of biology's studies in biology, no. 58. Edward Arnold, London, pp 15–45

Jordan CF (1985) Soils of the Amazon rainforest. In: Prance GT, Lovejoy TE (eds) Key environments: Amazonia. Pergamon Press, New York, pp 83–94

Kammesheidt L (1998) The role of tree sprouts in the restoration of stand structure and species diversity in tropical moist forest after slash-and-burn agriculture in Eastern Paraguay. Plant Ecol 139:155–165

Kanjunt C, Oberhauser U (1994) Successional forest development in abandoned swidden plots of Hmong, Karen, and Lisu ethnic groups. NAMSA Research Paper, Sam Muen-Highland Development Project, Thailand

Kappelle M, Geuze T, Leal ME, Cleef AM (1996) Successional age and forest structure in a Costa Rican upper montane *Quercus* forest. J Trop Ecol 12:681–698

Keen FGB (1978) Ecological relationships in a Hmong (Meo) Economy. In: Kunstadter P, Chapman EC, Sabhasri S (eds) Farmers in the forest: economic development and marginal agriculture in northern Thailand, East-West Center, University Press of Hawaii Honolulu, Hawaii pp 210–221

Kennard DK (2002) Secondary forest succession in a tropical dry forest: patterns of development across a 50-year chronosequence in lowland Bolivia. J Trop Ecol 18:53–66

Klanderud K, Mbolatiana HAH, Vololomboahangy MN, Radimbison MA, Roger E, Totland Ø, Rajeriarison C (2010) Recovery of plant species richness and composition after slash-and-burn agriculture in a tropical rainforest in Madagascar. Biodivers Conserv 19:187–204

Koutika LS, Sanginga N, Vanlauwe B, Weise S (2002) Chemical properties and soil organic matter assessment under fallow systems in the forest margins benchmark. Soil Biol Biochem 34(6):757–765

Kuzee M, Wideven S, De Haan T (1994) Secondary forest succession: analysis of structure and species composition of abandoned pastures in the Monteverde Cloud Forest Reserve, Costa

Rica. International Agricultural College Larenstein, Velp, and Agricultural University Wageningen, Wageningen

Lamb D (1980) Soil nutrient mineralization in secondary rainforest succession. Oecologia 47:257–263

Lee K (2002) Secondary succession in abandoned fields after shifting cultivation in Kangwon-Do, Korea. In: Lee D, Jin V (eds) Ecology of Korea, Bumwoo Publishing Company, Seoul, p 406

Lu D, Moran E, Mausel P (2002) Linking Amazonian secondary succession forest growth to soil properties. Land Degrad Dev 13:331–343

Martinez-Yrizar A, Burquez A, Maass M (2000) Structure and functioning of tropical deciduous forest in western Mexico. In: Robichaux RH, Yetman DA (eds) The tropical deciduous forest of Alamos. The University of Arizona Press Tucson, Arizona, pp 19–35

Méndez-Bahena A (1999) Sucesión secundaria de la selva húmeda y conservación de recursos naturales en Marqués de Comillas, Chiapas. M.Sc. Thesis, Departamento de Ecologia de los Recursos Naturales, Universidad Nacional Autónoma de México, Morelia, Mexico

Metzger JP (2003) Effects of slash-and-burn fallow periods on landscape structure. Environ Conserv 30(4):325–333

Miller PM, Kauffman JB (1998) Effects of slash and burn agriculture on species abundance and composition of a tropical deciduous forest. For Ecol Manag 103:191–201

Milleville P, Grouzis M, Razanaka S, Razafindrandimby J (2000) Sysèmes de culture sur abattis-brûlis et déterminisme de l'abandon cultural dans une zone semi-aride du sud-ouest de Madagascar. In: John Libbet Eurotext (eds) Actes du séminaire internationale, Dakar 13–16 avril 1999. Paris, pp 59–72

Mitja D, de Souza Miranda I, Velasquez E, Lavelle P (2008) Plant species richness and floristic composition changes along a rice-pasture sequence in subsistence farms of Brazilian Amazon, influence on the fallows biodiversity (Benfica, State of Pará). Agric Ecosyst Environ 124:72–84

Molina Colón S, Lugo AE (2006) Recovery of a subtropical dry forest after abandonment of different land uses. Biotropica 38:354–364

Moran EF, Brondizio ES, Mausel P (1994) Secondary succession. Natl Geogr Res Explor 10(4):458–476

Moran EF, Brondizio ES, Tucker J, Falesi IC, McCracken S (1996) Effects of soil fertility and land use on forest succession in Amazonia. Paper presented at the symposium ecological processes in lowland tropical secondary forests at the 1996 annual meeting of the ecological society of America, Providence, Rhode Island

Moran EF, Brondizio ES, Tucker JM, da Silva-Fosberg MC, McCracken S, Falesi I (2000a) Effects of soil fertility and land-use on forest succession in Amazonia. For Ecol Manag 139:93–108

Moran EF, Brondizio ES, Tucker J, Silva-Forsberg MC, Falesi IC, McCracken S (2000b) Strategies for Amazonian Forest Restoration: Evidence for Afforestation in Five Regions of the Brazilian Amazon. In: Anthony Hall (ed) Amazonia at the Crossroads: the challenge of sustainable development. Institute of Latin American Studies, London, pp. 129–149

Murphy PG, Lugo AE (1986a) Structure and Biomass of a Subtropical Dry Forest in Puerto Rico. Biotropica 18(2):89–96

Murphy PG, Lugo AE (1986b) Ecology of Tropical Dry Forest. Ann Rev Ecol Syst 17:67–88

Murhy PG, Lugo AE (1986) Ecology of tropical dry forest. Annu Rev Ecol Syst 17:89–96

Nakano K (1978) An ecological study of swidden agriculture at a village in Northern Thailand. Southeast Asian Study 16:411–445

Neba NE (2009) Cropping systems and post-cultivation vegetation successions: agro-ecosystems in Ndop. Cameroon. J Hum Ecol 27(1):27–33

Neil C, Melillo JM, Steudler PA, Cerri CC, de Moraes JFL, Piccolo MC, Brito M (1997) Soil carbon and nitrogen stocks following forest clearing for pasture in the southwestern Brazilian Amazon. Ecol Appl 7:1216–1225

Nyerges AE (1989) Coppice swidden fallows in tropical deciduous forest: biological, technological, and sociocultural determinants of secondary forest successions. Hum Ecol 17(4):379–400

Odum EP (1960) Organic production and turnover in old-filed succession. Ecology 18:243–253

Odum EP, Pinder JE III, Chrisliansen TA (1984) Nutrient losses from sandy soils during old-field succession. Am Midl Nat 111:148–154

Pickett STA, Collins SL, Armesto JJ (1987) Models, mechanisms and pathways of succession. Botanical Rev 53:335–371

Quesada M, Sanchez-Azofeifa GA, Alvarez-Anorve M, Stoner KE, Avila-Cabadilla L, Calvo-Alvarado J, Castillo A, Espírito-Santo MM, Fagundes M, Fernandes GW, Gamon J, Lopezaraiza-Mikel M, Lawrence D, Morellato LPC, Powers JS, Neves de FS, Rosas-Guerrero V, Sayago R, Sanchez-Montoya G (2009) Succession and management of tropical dry forests in the Americas: review and new perspectives. For Ecol Manag 258:1014–1024

Raharimalala O, Buttler A, Ramohavelo CD, Razanaka S, Sorg JP, Gobat JM (2010) Soil-vegetation patterns in secondary slash and burn successions in Central Menabe. Madag Agric Ecosyst Environ 139:150–158

Read L, Lawrence D (2003) Recovery of biomass following shifting cultivation in dry tropical forests of the Yucatan. Ecol Appl 13(1):85–97

Reiners WA, Bouwman AF, Parsons WFJ, Keller M (1994) Tropical rain forest conversion to pasture: changes in vegetation and soil properties. Ecol Appl 4:363–377

Rico-Gray V, García-Franco JG, Puch A, Sima P (1988) Composition and structure of a tropical dry forest in Yucantan, Mexico. J Ecol Environ Sci 4:21–29

Romero-Duque LP, Jaramillo VJ, Pérez-Jiménez A (2007) Structure and diversity of secondary tropical dry forests in Mexico, differing in their prior land-use history. For Ecol Manag 253:38–47

Rouw A (1993) Regeneration by sprouting in slash and burn rice cultivation. Taï rain forest, Côte d'Ivoire. J Tropical Ecol 9:387–408

Ruiz J, Fandiño MC, Chazdon RL (2005) Vegetation structure, composition, and species richness across a 56-year chronosequence of dry tropical forest on Providencia Island. Colombia, Biotropica 37(4):520–530

Saldarriaga JG, West DC, Tharp ML (1986) Forest succession in the upper Rio Negro of Colombia and Venezuela. Environmental sciences division publication no. 2694. Oak Ridge National Laboratory, Oak Ridge. p 164

Saldarriaga JG, West DC, Tharp ML, Uhl C (1988) Long-term chronosequence of forest succession in the upper Rio Negro of Colombia and Venezuela. J Ecol 76:938–958

Schmidt-Vogt D (1998) Defining degradation: the impacts of swidden on forests in Northern Thailand. Mt Res Devel 18(2):135–149

Schmidt-Vogt D (1999) Swidden farming and fallow vegetation in Northern Thailand, Geoecological Research, vol 8. Franz Steiner Verlag, Stuttgart

Segura G, Balvanera P, Durán E, Pérez A (2003) Tree community structure and stem mortality along a water availability gradient in a Mexican torpical dry forest. Plant Ecol 169:259–271

Silver WL, Scatena FN, Johnson AH, Siccama TG, Watt F (1996) At what temporal scales does disturbance affect below-ground nutrient pools? Biotropica 28:441–457

Styger E (2004) Fire-less alternatives to slash-and-burn agriculture (tavy) in the rainforest region of Madagascar, Dissertation, Department of Soil and Crop Sciences, Cornell University, Ithaca

Styger E, Rakotondramasy HM, Pfeffer MJ, Fernandes ECM, Bates DM, (2007) Influence of slash-and-burn farming practices on fallow succession and land degradation in the rainforest region of Madagascar. Agric Ecosyst Environ 119(3–4):257–269

Toky OP, Ramakrishnan PS (1983) Secondary succession following slash and burn agriculture in Northeastern India. II. Nutrient cycling. J Ecol 71:747–757

Tran P, Marincioni F, Shaw R (2010a) Catastrophic flood and forest cover change in the Huong river basin, central Viet Nam: A gap between common perceptions and facts. J Environ Manag 91(11):2186–2200

Tran VD, Osawa A, Nguyen TT (2010b) Recovery process of a mountain forest after shifting cultivation in Northwestern Vietnam. For Ecol Manag 259:1650–1659

Tran VD, Osawa A, Nguyen TT, Nguyen BV, Bui TH, Cam QK, Le TT, Diep XT (2011) Population changes of early successional forest species after shifting cultivation in Northwestern Vietnam. New Forest 41:247–262

Tucker JM, Brondizio ES, Morán EF (1998) Rates of forest regrowth in Eastern Amazonia: a comparison of Altamira and Bragantina regions. Pará State, Brazil, Interciencia 23(2):64–73

Uhl C (1987) Factors controlling succession following slash-and-burn agriculture in Amazonia. J Ecol 75:377–407

Uhl C, Clark K, Clark H, Murphy P (1981) Early plant succession after cutting and burning in the upper Rio Negro region of the Amazon basin. J Ecol 69:631–649

Uhl C, Jordan C, Clark K, Clark H, Herrera R (1982) Ecosystem recovery in Amazon caatinga forest after cutting, cutting and burning, and bulldozer clearing treatments. Oikos 38:313–320

Vieira ICG, Proctor J (2007) Mechanisms of plant regeneration during succession after shifting cultivation in eastern Amazonia. Plant Ecol 192:303–315

Voeks RA (1996) Tropical forest healers and habitat preference. Econ Botany 50(4):381–400

Walker AR (1975) Farmers in the Hills: upland peoples of Northern Thailand. Penerbit University Sains Malaysia, Chiang Mai

Weaver PL, Birdsey RA, Lugo AE (1987) Soil organic matter in secondary forests of Puerto Rico. Biotropica 19:17–23

Werner P (1984) Changes in soil properties during tropical wet forest succession in Costa Rica. Biotropica 16:43–50

Whigham, DF, Zugasty Towle P, Cabrera Cano E, O'Neill J, Ley, E (1990) The effect of annual variation in precipitation on growth and litter production in a tropical dry forest in the Yucatan of Mexico. J Trop Ecol 31(2):23–34

Williams-Linera G (1983) Biomass and nutrient content in two successional stages of tropical wet forest in Uxpanapa, Mexico. Biotropica 15:275–284

Williams-Linera G, Lorea F (2009) Tree species diversity driven by environmental and anthropogenic factors in tropical dry forest fragments of central Veracruz, Mexico. Biodivers Conserv 18:3269–3293

Williams-Linera G, Alvarez-Aquino C, Hernández-Ascención E, María T (2011) Early successional sites and the recovery of vegetation structure and tree species of the tropical dry forest in Veracruz. Mexico, New Forests 42:131–148

Chapter 6
Conclusions

Keywords Rate of recovery · Successional process · Forest structure · Recovery rate · Biomass accumulation rate · Forest functions

This book has reviewed case studies on ecological succession following shifting cultivation from different countries and forest types. Changes in soil conditions and different aspects of vegetation were discussed. These aspects include species richness, species diversity, aboveground biomass, basal area, tree/canopy height, plant density and species composition. It would be hard to generalize from all of the results as they are very site specific, and comparison across studies is intractable as the number of fallow stands, plot size, and methods used to evaluate vegetation changes along succession were different between studies. Although it is difficult to draw conclusions about the vegetation changes along succession, forest structures such as basal area and canopy height illustrate a relatively obvious increasing trend regardless of location and type of forest. On the other hand, changes in species richness, species diversity and species composition tend to display great variation among studies.

The intermediate disturbance hypothesis was found to be valid in some of the studies only, and not all results agreed with this prediction. Similarly, both the 'initial floristic composition' (IFC) and the 'relay floristics' model (Chap. 4) seem unable to fully explain the changes in species composition of many studies. Moreover, a high variation exists in the time needed for fallows to reach old-growth forest levels for measures of forest structure and species composition. Some examples we outlined demonstrated that a longer time is needed for species composition to attain a similar level to that of the old-growth forest, compared to other measures of forest structure (e.g. Aweto 1981a; Peña-Claros 2003; Kammesheidt 1998). However, not all case studies agreed with this conclusion.

Some authors claim that dry forests tend to recover more rapidly than moist forests to old-growth forest levels (Ewel 1977; Murhy and Lugo 1986; Segura et al. 2003). This is thought to be because of lower canopy height and simpler structure

(Ewel 1980; Murhy and Lugo 1986). However, according to our review of the literature, dry forests do not necessarily have a higher rate of recovery or resilience than moist forests. Our review is thus in agreement with results generated by Quesada et al. (2009). The number of studies on dry forests is significantly less than that conducted on moist forests. There is certainly a need to pay more attention to the ecological succession following shifting cultivation in dry forests.

A number of researchers pointed out that the succession process is driven by a lot of factors such as land use history, climate, soil, etc., which are all interrelated (Guariguata and Ostertag 2001) and of equal importance (Pickett and White 1985; Redzic 2000; El-Sheikh 2005). As mentioned by Lebrija-Trejos et al. (2008), the time needed for forest structure to return to pre-existing mature forest conditions depends on different factors, like the quality of the old-growth forests used for comparison, the land use history and the sampling criteria used. Shifting culti-vation intensity and frequency of burning might also contribute to the alteration of the successional process, resulting in great divergences among places. Due to these anthropogenic disturbances and locational variations, succession tends to be site specific and difficult to predict.

In addition, how these variables impact the regeneration of secondary forest is still unclear. Studies that we reviewed here suggested that a higher frequency of burning might destroy the soil seed bank and ability to sprout. Yet, how important seed availability and sprouting ability are in different stages of succession for forest regeneration, especially in the long-run, is still unclear. Some researchers pointed out that the longer farming duration tends to retard secondary forest succession, while the higher intensity of land clearing might reduce coppicing. The importance of coppicing on forest regeneration was shown in some studies, but similarly, any conclusion on how it impacts the rate of forest regrowth in the long-run necessitates more research.

Another factor that might influence succession are soil properties, as suggested by a number of researches. Contrasting opinions have been presented here. For example, many studies concluded that biomass accumulation and forest structure such as basal area, plant density and tree height are greater on fallows with more fertile soil. In contrast, some studies found that no significant differences in species richness existed between plots with fertile soils versus less fertile soils. Thus, how soil fertility influences the rate of succession and vegetation growth would be important avenues of future research. Due to the complex interrelationship between the variables mentioned above, the successional process is hard to predict and to generalize. To better understand the successional process, studies with longer chronosequences are needed. Our review found that most researches examined sites with less than 50 years of fallow. A study like that of Saldarriaga et al. (1988), who examined an 80-year chronosequence, are extremely rare. Longer chronosequence studies would be beneficial to the understanding of sec-ondary forest succession and how vegetation and soil changes along the process.

How does one determine the rate of recovery? What should be the indicator used to determine the growth rate of secondary forest? To evaluate the successional process of the fallows, most studies compare the vegetation on the fallows to that in the local old-growth forest (see also Schmidt-Vogt 1998). However, a number of studies pointed out that it is difficult to make a comparison with old-growth forests as there might not be undisturbed forests left in some regions (Schmidt-Vogt 1998; Chazdon et al. 2007; Budowski 1970; Gentry 1991; Kennard 2002). Furthermore, some of these forests might be late secondary rather than primary forests (Brown and Lugo 1990). Kennard (2002) suggested that the recovery rate might be overestimated if the control forests had been disturbed. Thus, while it is still important to use old-growth forest as a control when examining the secondary forest regeneration process, they should be carefully picked to avoid bias. Also, to determine the rate of recovery of fallow fields, merely focusing on how the forest structures can attain old-growth forest levels is insufficient. Some studies found that particular characteristics, such as species richness, species diversity, basal area and plant density, were even greater on fallows than in old-growth forests. Furthermore, according to Guariguata and Ostertag (2001), forest functions such as nutrient recycling and net primary productivity should also be considered, but are seldom mentioned.

In this book, we also reviewed studies that examine how slash-and-burn cycles, shifting cultivation intensity, soil properties and types of forest impact the successional process. However, there are more variables, such as distance to mature forests, which are proven to affect the rate of vegetation regeneration on the fallows (Johnson et al. 2000; Moran et al. 2000; Zarin et al. 2001; Chazdon 2003). The interrelationship between these factors also warrants future attention, as few studies have been done to investigate how these factors interact to alter the successional process. These complicated interactions could help improve our understanding of the management of fallow fields, so as to prevent land from degrading. Another topic that needs more focus is the soil-vegetation relation. In most studies on ecological succession, researchers focus only on how the vegetation changes, but seldom discuss how soil fertility changes. As demonstrated in Lu et al. (2002), there is a clear interrelation between soil and vegetation, in which soil fertility affects the biomass accumulation rate. More studies should be carried out to further understand how fallow length and soil fertility are related, and, on the other hand, if less trees are available in shorter fallows, how the abundance of trees relates to soil fertility. As mentioned earlier, Guariguata and Ostertag (2001) pointed out that forest functions should be discussed more in studies on secondary forest regeneration. How soil nutrient recycling occurs and how it impacts vegetation growth on these agricultural plots should serve as one of the indicators to evaluate forest recovery rates.

Another area that should be the focus of more attention is the models used to predict how species composition changes along the successional trajectory. Breugel et al. (2007) concluded that their study only partly agreed with the IFC model. Thus, they concluded that a new approach is needed to explain the high variation in species composition among fallows. From our review, it is clear that a great variability of changes in species composition exists among studies, and even among fallows that are the same ages. Both the 'IFC' and the 'rely floristic' models might not be able to fully reflect and predict the differences between fallow fields. Therefore, new models or approaches might be needed.

To conclude, our review of case studies from different parts of the world found that species richness, species density, aboveground biomass, basal area, tree height, plant density and species composition, all show great variations during succession, and are very site specific. To facilitate the understanding of ecological succession following shifting cultivation, we advocate that a larger focus is put on the interrelationship between different environmental variables. Forest functions should also be considered when evaluating forest recovery rates, as focusing exclusively on forest structures and species richness, diversity and composition might not be sufficient to understand the ecological succession on fallows.

References

Aweto AO (1981) Secondary succession and soil fertility restoration in south-western Nigeria, I, Succession. J Ecol 69:601–607

Breugel MV, Bongers F, Martínez-Ramos M (2007) Species dynamics during early secondary forest succession: recruitment, mortality and species turnover. Biotropica 35:610–619

Brown S, Lugo AE (1990) Tropical secondary forests. J Trop Ecol 6:1–32

Budowski G (1970) The distinction between old secondary and climax species in tropical Central American lowland forests. Trop Ecol 2:44–48

Chazdon RL (2003) Tropical forest recovery: legacies of human impact and natural disturbances. Perspect Plant Ecol Evol Syst 6:51–71

Chazdon RL, Letcher SG, van Breugel M, Martínez-Ramos M, Bongers F, Finegan B (2007) Rates of change in tree communities of secondary Neotropical forests following major disturbances. Philos Trans R Soc B 362:273–289

El-Sheikh MA (2005) Plant succession on abandoned fields after 25 years of shifting cultivation in Assuit, Egypt. J Arid Environ 61:461–481

Ewel J (1977) Differences between wet and dry successional tropical ecosystems. Geogr-Ecol-Trop 1:103–117

Ewel JJ (1980) Tropical succession: manifold routes to maturity. Biotropica 12:2–9

Gentry AH (1991) The distribution and evolution of climbing plants. In: Bullock SH, Mooney HA (eds) The biology of vines. Cambridge University Press, Cambridge, pp 3–42

Guariguata MR, Ostertag R (2001) Neotropical secondary forest succession: changes in structural and functional characteristics. For Ecol Manage 148:185–206

Johnson CM, Zarin DJ, Johnson AH (2000) Post-disturbance aboveground biomass accumulation in global secondary forests. Ecology 81:1395–1401

Kammesheidt L (1998) The role of tree sprouts in the restoration of stand structure and species diversity in tropical moist forest after slash-and-burn agriculture in Eastern Paraguay. Plant Ecol 139:155–165

Kennard DK (2002) Secondary forest succession in a tropical dry forest: patterns of development across a 50-year chronosequence in lowland Bolivia. J Trop Ecol 18:53–66

Lebrija-Trejos E, Bongers F, Pérez-García EA, Meave JA (2008) Successional change and resilience of a very dry tropical deciduous forest following shifting agriculture. Biotropica 40:422–431

Lu D, Moran E, Mausel P (2002) Linking Amazonian secondary succession forest growth to soil properties. Land Degrad Dev 13:331–343

Moran EF, Brondizio E, Tucker JM, da Silva-Fosberg MC, McCracken S, Falesi I (2000) Effects of soil fertility and land-use on forest succession in Amazonia. Forest Ecol Manag 139:93–108

Murhy PG, Lugo AE (1986) Ecology of tropical dry forest. Annu Rev Ecol Syst 17:89–96

Peña-Claros M (2003) Changes in forest structure and species composition during secondary forest succession in the Bolivian Amazon. Biotropica 35(4):450–461

Pickett STA, White PS (1985) Patch dynamics: a synthesis. In: Pickett STA, White PS (eds) The ecology of natural disturbance and patch dynamics. Academic Press, New York

Quesada M, Sanchez-Azofeifa GA, Alvarez-Anorve M, Stoner KE, Avila-Cabadilla L, Calvo-Alvarado J, Castillo A, Espírito-Santo MM, Fagundes M, Fernandes GW, Gamon J, Lopezaraiza-Mikel M, Lawrence D, Morellato LPC, Powers JS, Neves F de S, Rosas-Guerrero V, Sayago R, Sanchez-Montoya G (2009) Succession and management of tropical dry forests in the Americas: review and new perspectives. Forest Ecol Manag 258:1014–1024

Redzic S (2000) Patterns of succession of xerophylous vegetation on Balkans. In: White PS, Mucina L, Lepš J (eds) Vegetation science in retrospect and perspective. Opulus press, Uppsala, pp 76–79

Saldarriaga JG, West DC, Tharp ML, Uhl C (1988) Long-term chronosequence of forest succession in the upper Rio Negro of Colombia and Venezuela. J Ecol 76:938–958

Schmidt-Vogt D (1998) Defining degradation: the impacts of swidden on forests in Northern Thailand. Mt Res Dev 18(2):135–149

Segura G, Balvanera P, Durán E, Pérez A (2003) Tree community structure and stem mortality along a water availability gradient in a Mexican torpical dry forest. Plant Ecol 169:259–271

Zarin DJ, Ducey MJ, Tucker JM, Salas WA (2001) Potential biomass accumulation in Amazonian regrowth forests. Ecosystems 4:658–668